AF764812

ADVANCES IN MARINE BIOLOGY

VOLUME 1

ADVANCES IN MARINE BIOLOGY

Additional books in this series can be found on Nova's website
under the Series tab.

Additional e-books in this series can be found on Nova's website
under the e-book tab.

ADVANCES IN MARINE BIOLOGY

VOLUME 1

ADAM KOVÁCS
AND
PATRIK NAGY
EDITORS

nova publishers
New York

Copyright © 2016 by Nova Science Publishers, Inc.

All rights reserved. No part of this book may be reproduced, stored in a retrieval system or transmitted in any form or by any means: electronic, electrostatic, magnetic, tape, mechanical photocopying, recording or otherwise without the written permission of the Publisher.

We have partnered with Copyright Clearance Center to make it easy for you to obtain permissions to reuse content from this publication. Simply navigate to this publication's page on Nova's website and locate the "Get Permission" button below the title description. This button is linked directly to the title's permission page on copyright.com. Alternatively, you can visit copyright.com and search by title, ISBN, or ISSN.

For further questions about using the service on copyright.com, please contact:
Copyright Clearance Center
Phone: +1-(978) 750-8400 Fax: +1-(978) 750-4470 E-mail: info@copyright.com.

NOTICE TO THE READER

The Publisher has taken reasonable care in the preparation of this book, but makes no expressed or implied warranty of any kind and assumes no responsibility for any errors or omissions. No liability is assumed for incidental or consequential damages in connection with or arising out of information contained in this book. The Publisher shall not be liable for any special, consequential, or exemplary damages resulting, in whole or in part, from the readers' use of, or reliance upon, this material. Any parts of this book based on government reports are so indicated and copyright is claimed for those parts to the extent applicable to compilations of such works.

Independent verification should be sought for any data, advice or recommendations contained in this book. In addition, no responsibility is assumed by the publisher for any injury and/or damage to persons or property arising from any methods, products, instructions, ideas or otherwise contained in this publication.

This publication is designed to provide accurate and authoritative information with regard to the subject matter covered herein. It is sold with the clear understanding that the Publisher is not engaged in rendering legal or any other professional services. If legal or any other expert assistance is required, the services of a competent person should be sought. FROM A DECLARATION OF PARTICIPANTS JOINTLY ADOPTED BY A COMMITTEE OF THE AMERICAN BAR ASSOCIATION AND A COMMITTEE OF PUBLISHERS.

Additional color graphics may be available in the e-book version of this book.

Library of Congress Cataloging-in-Publication Data

ISBN: 978-1-53610-131-7

Published by Nova Science Publishers, Inc. † New York

CONTENTS

PREFACE

Advances in Marine Biology" presents original research results on the leading edge of marine biology discovery. Each article has been carefully selected in an attempt to present substantial research results across a broad spectrum. In this volume, the first chapter reviews the influenza viruses in marine mammals. Chapter Two presents an overview of the development and study of marine sponge's bioactive compounds for anti-inflammatory activity for last four years i.e., 2012-2016. Chapter Three provides an updated review of the helminth fauna of cetaceans. Chapter Four discusses the effect of clays on the removal of okadaic acid from seawater. Chapter Five focuses on biomarker responses in black sea scorpian fish *Scorpaena porcus* to anthropogenic impact. Chapter Six provides an overview of trophic ecology of the shortfin mako *Isurus oxyrinchus* (Lamniformes: Lamnidae) in the Eastern Pacific ocean. Chapter Seven examines a fishery encompassing two of the Spanish fleet's target species that exhibit a significant ecological predatory interdependence: the blue whiting and the southern stock of European hake.

Chapter 1 - Marine mammals may have an important influence on the marine communities and their conservation has a great impact on balance in the ocean's ecosystems. Marine mammals mortality due to infectious disease and algal biotoxins are increasing worldwide. Morbilliviruses and influenza viruses are two major viral infections that caused several outbreaks with high morbidity and mortality among marine mammals. Influenza viruses are among few zoonotic pathogens known to have caused infections in marine mammals. Direct transmission of influenza A and B viruses from humans to seals and vice versa, and mass die-off of seals due to influenza virus infections in several occasions accentuates their importance for the conservation of marine mammals population.

Chapter 2 - This chapter presents an overview of the development and study of marine sponge's bioactive compounds for anti-inflammatory activity for last four years i.e., 2012-2016. Marine pharmacology during 2009–2011 remained a global enterprise contributing to the preclinical pharmacology of 262 marine compounds which are in pharmaceutical pipeline. There is no updated review for the studies done on marine sponge's compounds showing anti-inflammatory activity after 2011. This chapter is a sincere effort to present a systematic review of the preclinical pharmacology of marine sponge and its associated microbes/symbionts compounds for potential activity in inflammation.

Chapter 3 - Parasitism is an extraordinary life-strategy that largely influences that of most free-living organisms, including cetaceans. Parasites of cetaceans have a positive intrinsic and instrumental value, but also a potential pathogenic impact on host populations and, therefore, they should be considered in any biodiversity or conservation program. Yet, a starting point for understanding the relationships between cetaceans and their parasites is to have a detailed account on parasite diversity. The knowledge of the parasite fauna of cetaceans goes back to the time when commercial whaling was allowed and material for scientific purposes was available. Over the last decades, however, parasites have been obtained mainly from stranded or by-caught animals, and regular updates of host-parasite lists have been published until the 1990's. In this chapter the authors provide an updated review of the helminth fauna of cetaceans. The data here presented constitutes a baseline for future diversity surveys and a way to encourage researchers towards a greater awareness about the biological importance of the helminth fauna of cetaceans. Out of the 90 species currently recognized in the order Cetacea, 72 have hitherto been examined for helminths. The family Balaenopteridae among baleen whales and the family Delphinidae among toothed whales harbor the most diverse helminth fauna, which is partly accounted for by a higher sampling effort. In contrast, the helminth fauna of the beaked whales (family Ziphiidae) is the least known since only 9 out of the 22 species in the family have been examined for parasites. Currently, there are 174 helminth species reported in cetaceans, from which nematodes is the most speciose group (62 spp.), followed by digeneans (54 spp.), cestodes (38 spp.) and acanthocephalans (20 spp.). However, 20 (11.5%) of these species represent accidental infections that occur because cetaceans and other marine vertebrates (i.e., pinnipeds and marine birds) share common prey. The families Anisakidae, Pseudaliidae and Tetrameridae (Nematoda), Brachycladiidae, Notocotylidae, Heterophyidae and Brauninidae (Digenea), Diphyllobothriidae,

Phyllobothriidae and Tetrabothriidae (Cestoda), and Polymorphidae (Acanthocephala) have a major species-representation in the helminth fauna of cetaceans.

Chapter 4 - The use of clays for mitigating the effects of toxic episodes causing continuous shellfish harvesting closures owing to lipophilic toxins remains unexplored. This strategy is promising for clearance of microalgae through flocculation, but few studies evaluate toxins elimination. This study presents preliminary data on the withdrawal of okadaic acid, testing different clays in seawater, in laboratory conditions. Clays were assayed based on the potential effectiveness in terms of specific surface, particle size and colloidal properties, as well as their local abundance: three kinds of bentonites (one containing sodium bentonite particles <70 μm, other, 83% < 2 μm, and bentonite mainly composed of muscovite, particles >45 μm) and kaolinites of different origins and grain size (69% < 2 μm and 57% < 2 μm). Bentonite (83% < 2 μm) had the best toxin removal ability in the authors' study (68%), while kaolinites with longer decantation times (69% < 2 μm) were the least effective material. The particles with the highest size (bentonite >45 μm) showed a lower efficiency, probably related to the fast decantation process. Altogether, clay applications offer an opportunity for toxin removal that should be explored in depth.

Chapter 5 - Long-term and large-scale monitoring studies indicate the changes of anthropogenic impact on the water ecosystems, which can be chronically stressed by multiple environmental factors. Fish are very sensitive to anthropogenic pollution, and some of them may be tested as biomonitors for the evaluation of the ecological status and risk assessment of marine environments. Fish biochemical parameters could be directly related to the area where the fish were collected. In polluted sites the exposure of fish to xenobiotics results to interaction between these chemicals and biological systems which may give elevation to biochemical and physiological damage or/and adaptive mechanisms via the induction of defense immune and antioxidant systems. Biochemical and physiological parameters are used as biomarkers for contaminants and could be applied for the evaluation of environmental stress and its after-effects in fish. Biomarkers exposure to environmental stressors vary widely depending on the type of anthropogenic activity involved. Sewage and chemical pollution from industrial, agricultural, maritime transport and domestic effluents are the main sources of pollution in Black Sea ecosystem. Coastal waters are the main recipients of discharges and combine effects of the various kinds of contaminants result dramatic ecological consequences such as eutrophication, biodiversity loss, elimination of some species, worsening of their health and decline the population size. Biochemical biomarkers variables seem to be useful indicators of the health status of fish and their habitats in monitoring studies. Scorpion fish (*Scorpaena porcus*) is among the most common fish species in Black Sea coastal waters, and it was selected as biomonitor species. The biomarker response in scorpion fish caught in several Sevastopol bays characterizing different levels of pollution was studied related to anthropogenic impact. Selected biomarkers of exposure and effect, including parameters of oxidative stress and oxidative damage, were good tools for the evaluation of fish health and ecological status of their habitats.

Chapter 6 - Shortfin mako sharks *Isurus oxyrinchus* are important organisms in the pelagic ecosystem. In spite of their vulnerable status, they are captured by fishermen in the Eastern Pacific Ocean. The proper management and conservation of this population in the Pacific is critical and thus, more biological knowledge of this species is needed. In this work, the trophic habits and consumption of key prey by *I. oxyrinchus* were studied through stomach content analysis of samples collected off the coast of Ecuador, and a literature review of its feeding habits was conducted for the eastern Pacific. The index of relative importance of prey items, and the consumption of key prey were calculated for the three areas. The diet was composed of 78 prey taxa, and cephalopods were the main prey items in the north and central areas, with the jumbo squid *Dosidicus gigas* being the most important prey. For the southern area, *I. oxyrinchus* fed on fish with the bigeye cigarfish *Cubiceps pauciradiatus* being the most important prey species. The trophic niche of *I. oxyrinchus* was composed of squids and fish as the main prey items. The niche breadth for the three areas was reduced, and the trophic level calculated with stomach contents and stable isotopes of $\delta^{15}N$ showed values of 3.68 to 4.44 suggesting that this shark is a tertiary predator in the ecosystem. The jumbo squid *D. gigas* and fish of the family Scombridae were described as key species in the diet of *I. oxyrinchus*. The consumption estimate from 2000 to 2013 was 39 128 tons, with 55.8% of this consumption represented by *D. gigas*. These results demonstrate that *I. oxyrinchus* is an active and top predator in the trophic webs of the Eastern Pacific, with opportunistic feeding habits on the more available and abundant prey species.

Chapter 7 - Ecological interactions amongst fish stocks can play a dominant role in how marine ecosystems are structured, and they can have a significant impact on the populations' growth dynamics. The analytical framework for proposing management guidelines has been developed based on bio-economic models that only account for a single species, both in terms of biological and well as economic aspects, and this diminishes the importance of interactions among the various marine community components. Bearing in mind the implications of the ecological dependencies that are created in multi-species fisheries, that bio-economic model can be extended. A better understanding of trophic relationships among fish stocks enables the development of management models for multi-species fisheries—that is, for most European fisheries. This chapter examines a fishery encompassing two of the Spanish fleet's target species that exhibit a significant ecological predatory interdependence: the blue whiting and the southern stock of European hake. Such interactions should be considered when establishing the Total Allowable Catch (TAC) because the catch level for one species will alter populations of the other, thereby affecting long-run sustainability of commercial fishing and of the marine environment. Within this fishery, both the catch and the spawning biomass for each species are trending downward. The trophic interaction between predator (hake) and prey (blue whiting) is included into the management problem of jointly determining fishing quotas for each species. The estimations indicate that increases in blue whiting biomass (prey) have generated a positive effect on hake biomass along that period 1988-2014, it is represented by a plus sign in the parameter for interdependence in the hake growth function. However, increases in hake biomass (predator) lead to a reduction in blue whiting biomass and this is represented by a minus sign in the parameter for interdependence in the blue whiting growth function. These results are consistent with a relationship of interdependence predator-prey. Incorporating this interaction into stock dynamics for both species and applying optimum control theory, the authors obtain the optimal spawning stock biomass values for both species and the corresponding catch values. The results show catch levels are lower than prevailing EU-determined levels for both species. In addition, as the social discount rate increases and, therefore society's preferences in the near future, the biomass of both species decreases, while catches or potential landings increase.

In: Advances in Marine Biology. Volume 1 ISBN: 978-1-53610-131-7
Editors: Adam Kovács and Patrik Nagy © 2016 Nova Science Publishers, Inc.

Chapter 1

INFLUENZA VIRUSES: A THREAT TO MARINE MAMMALS POPULATIONS

Sasan Fereidouni *
Research Institute of Wildlife Ecology,
University of Veterinary Medicine, Vienna, Austria

ABSTRACT

Marine mammals may have an important influence on the marine communities and their conservation has a great impact on balance in the ocean's ecosystems. Marine mammals mortality due to infectious disease and algal biotoxins are increasing worldwide. Morbilliviruses and influenza viruses are two major viral infections that caused several outbreaks with high morbidity and mortality among marine mammals. Influenza viruses are among few zoonotic pathogens known to have caused infections in marine mammals. Direct transmission of influenza A and B viruses from humans to seals and vice versa, and mass die-off of seals due to influenza virus infections in several occasions accentuates their importance for the conservation of marine mammals population.

Keywords: marine mammals, influenza viruses, zoonotic diseases

---

* Corresponding author: Research Institute of Wildlife Ecology, University of Veterinary Medicine, Vienna, Austria. Email: sasan.fereidouni@vetmeduni.ac.at.

INTRODUCTION

Marine mammals are an important part of the marine communities and their conservation has a great impact on balance in the ocean's ecosystems. Marine mammals face a wide range of threats from human activities, hunting, climate change, pollution, diseases, and habitat degradation. Marine mammals mortality due to infectious disease and algal biotoxins are increasing worldwide.

Various disease outbreaks and mass mortality events, some involving hundreds to thousands of animals, affected the marine mammal populations. Mass mortality events due to outbreaks of phocine distemper virus (PDV) infections occurred in European harbour seals (Phoca vitulina) in 1988 with more than 23000 and again in 2002 with more than 30000 dead cases (Härkönen et al., 2006; Kreutzer et al., 2008). Outbreaks of dolphin morbillivirus infections in bottlenose dolphins along the Atlantic coast accounted for 750 dead dolphins in 1987-1988 (Lipscomp et al., 1994; Taubenberger, et al., 1996) and again in 2013-2015 with approximately 2000 dead cases. In the Mediterranean Sea also dolphin morbilliviruses caused striped dolphins die-off in 1990-1992 with more than 1000 found carcasses (Aguilar and Raga, 1993), and again in 2006-2007 with approximately 100 striped dolphins and additional 60 long-finned pilot whales in the Spanish Mediterranean coast (Fernández et al., 2008). Morbilliviruses also killed Baikal seals in Lake Baikal in 1988 (Mamaev et al., 1996) and thousands of Caspian seals in the Caspian Sea in 2000 (Kennedy et al., 2000).

The importance of marine mammals as hosts or carriers of potential zoonotic pathogens such as influenza viruses is not well understood. The fact that influenza viruses are some of the few zoonotic pathogens known to have caused infections in marine mammals and evidence for direct transmission of influenza viruses from seals to humans and vice versa, underlines the importance of investigation and monitoring of influenza viruses in marine mammal populations (Fereidouni et al., 2014).

INFLUENZA OUTBREAKS IN SEAL AND WHALE POPULATIONS

Our knowledge of disease outbreaks in seal populations is restricted to how efficient are reporting systems in the countries that are close to marine mammals communities. Therefore the real rate of infection most probably be higher than we expect. Influenza viruses were partly isolated from marine mammals during outbreaks with clinical symptoms and mortality, and partly just during conventional monitoring studies from apparently healthy animals (Fereidouni et al., 2014). The first reported case of influenza virus isolation from marine mammals dates back to 1975/76. In this study, tissue samples were collected from 72 whales hunted in the South Pacific and influenza viruses were isolated from 13 lung and one liver specimens. Only detailed data on one H1N3 subtype that isolated probably from a striped whale (family Balaenopteridae) was reported (Lvov et al. 1978).

The first reported case of influenza virus induced mortality in marine mammals dates back to December 1979 (Figure 1). Approximately 500 mainly juvenile harbour seals (Phoca vitulina) carcasses were found along the north-eastern coast of the United States. The causal virus was identified as an influenza A virus of the H7N7 subtype (Webster et al. 1981a; Lang et al. 1981). This epizootic infection affected at least 20% of the local seal population and some animals developed severe acute haemorrhagic viral pneumonia (Geraci et al. 1982). The causative virus was antigenically similar to an avian H7N7 virus. However, as its biological properties was more like a mammalian strain, the virus was probably adapted to mammalian hosts during replication in seals (Webster et al. 1981a; Kida et al. 1982; Naeve and Webster, 1983; Callan et al. 1995).

The next reported influenza outbreak occurred in the same location in Massachusetts, from June 1982 to August 1983 (Figure 1). Approximately 60 harbour seals died due to the infection with an H4N5 influenza A virus. The main pathological findings was pneumonia and the causative virus was isolated from the lungs and brains of dead seals (Hinshaw et al. 1984). This outbreak focused the scientific attention on the role of marine mammals in the ecology and epidemiology of influenza viruses in nature (Webster et al. 1992).

In 1984, two influenza A viruses (H13N2 and H13N9) were isolated simultaneously from the hilar lymph nodes and lungs of a sick long-finned pilot whale (Globicephala melas) near Portland, Maine, USA. In post-mortem examination, pathological findings included a large hilar lymph node and haemorrhagic lungs (Hinshaw et al. 1986). It was speculated that these H13 influenza viruses had probably been related with two mass stranding of long-finned pilot whales (97 whales in October, 23 whales in November 1984) along the New England coast (Hinshaw et al. 1986; Van Bressem et al. 1999; Waltzek et al. 2012).

In 1991, two influenza viruses of the H4N6 subtype were isolated from the lung tissue of two dead harbour seals. Pathological lesions consistent with influenza virus infection, such as acute interstitial and/or haemorrhagic pneumonia and subcutaneous emphysema were observed. There was no indication for an extensive outbreak in that period along the New England coast, and the viruses were found within the framework of an surveillance study (Callan et al. 1995).

Again in September 1991 to April 1992, an increase in the number of stranded seals along Cape Cod was reported. A severe epizootic of viral pneumonia in the seal population was not observed, but pathologic lesions in respiratory system, including acute interstitial pneumonia and subcutaneous emphysema were found in dead seals. For the first time influenza viruses of the H3N3 subtype were isolated from the harbour seals (Callan et al. 1995).

An influenza B virus was isolated in 1999 from a throat swab of a juvenile seal in a rehabilitation centre in the Netherlands. Phylogenetic analysis of the HA gene showed a close relation to viruses which had been circulating in humans in 1995 (Osterhaus et al. 2000). No influenza virus infection-specific clinical symptom was reported in that juvenile seal. A retrospective study of seal serum samples, collected before and after 1995 indicated the introduction of influenza B virus into the seal population in 1995. Less than 2% of sera collected after 1995 tested positive for antibodies to the influenza B virus, which indicated low incidence of infection (Fouchier et al. 2001).

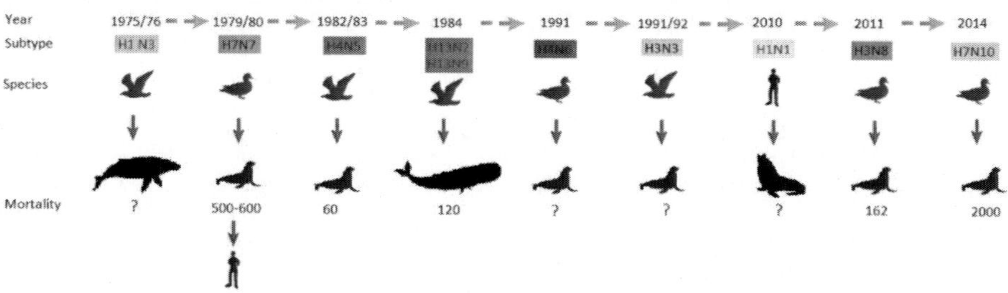

Figure 1. Influenza virus outbreaks/infections reported during 1975-2015. Arrows indicate known transmission routes between each species based on reported cases, phylogenetic analysis and experimental studies.

Following the human pandemic caused by a new H1N1 influenza virus (H1N1pdm09), nasals swabs were collected from 42 female free-ranging Northern elephant seals (Mirounga angustirostris) in a surveillance study along California's central coast in April 2010 (Goldstein et al. 2013). Two swab samples tested positive and the isolated H1N1 viruses showed an homology greater than 99% to influenza viruses isolated from humans in 2009. Serological analyses of serum samples collected between January 2010 and April 2011 indicated a recent introduction of the virus into the local seal population in April 2011. No mortality was seen in that seal population during the period of study (Goldstein et al. 2013).

In September - December 2011 higher mortality of harbour seals compared to previous years was reported from the New England coast. An avian influenza virus of the H3N8 subtype was isolated from several tissues of five out of 162 dead harbour seals. The main pathological findings were pneumonia and ulcerations of the skin and oral mucosa.

The first reported mass mortality of harbour seals in Europe due to influenza virus infections occurred in the Baltic Sea in 2014. From March to October, 425 carcasses were detected at the west coast of Sweden (Zohari et al., 2014), 152 carcasses on the small island of Anholt in Denmark (Krog et al., 2015), and approximately 1,400 carcasses in the north coastal waters of Germany (Bodewes et al., 2015). Avian influenza A virus (H10N7) was detected in the lungs of affected animals in all three occasions with close genetic relatedness. This subtype had not been reported in seals until 2014 and was associated with the highest mortality ever has been reported in seals due to influenza viruses. In addition, preliminary analysis of the haemagglutinin gene (HA) sequence of the causative influenza virus suggested the presence of molecular determinants that indicated mammalian adaptation (Bodewes et al., 2015).

Several serological studies have been carried out since 1978 to estimate the prevalence of influenza virus infections among marine mammals. Different methods such as haemagglutination inhibition (HI), indirect enzyme linked immunosorbent assay (ELISA) and competitive ELISA were used. Each method had its intrinsic sensitivity and specificity, positive predictive values (PPV) and negative predictive value (NPV); however, in most cases these methods were never validated for marine mammal serum samples (Fereidouni et al., 2014). Nevertheless, positive serum samples were found for harbour seals and sea lions (de Boer et al., 1990), harp and hooded seals (Stuen et al., 1994), ringed seal and beluga whales (Austin and Webster, 1993; Nielsen et al., 2001), Caspian seals, common minke whales and Dall's porpoise (Ohshi et al., 2002, 2006), Pacific walrus and bearded seals (Calle et al., 2002, 2008), Kuril harbour seals (Fujii et al., 2007), and fur seals (Blanc et al., 2009). In addition, influenza B antibodies were found in harbour and gray seals (Fouchier et al., 2001; Bodewes et al., 2013).

Genetic Background of Marine Mammals Influenza Viruses

Several influenza viruses have been isolated from marine mammals since the 1970s and sequence data of one to eight genome segments of them were obtained using different sequencing methods. Phylogenetic, genetic and/or antigenic analyses using different methods has been indicated that the majority of marine mammal influenza viruses showed high genetic relatedness with those viruses reported from wild birds.

Probably the first phylogenetic analysis for marine mammals influenza viruses was performed in 1982 using the full sequenced haemagglutinin gene of H7N7 virus and its association with avian H7 viruses was proved (Kida et al. 1982; Naeve and Webster, 1983).

Another phylogenetic analysis was performed in 1990, to make a comparison between the nucleoprotein (NP) gene sequence of three influenza viruses, H1N3, H13N2 and H7N7 isolated from stripped whales, long-finned pilot whales and harbour seals, respectively. The results indicated again high genetic relatedness between marine mammal viruses and those reported from wild birds (Mandler et al. 1990).

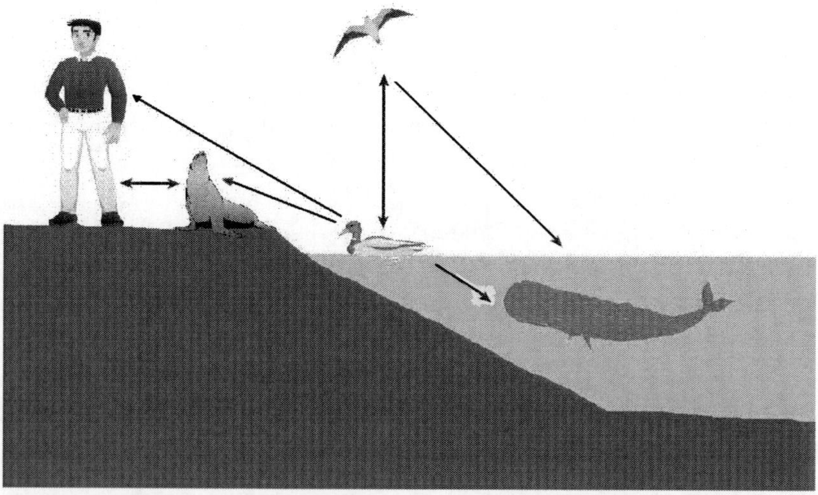

Figure 2. Inter-species transmission of influenza viruses between marine mammals and other species based on genetic relatedness of the isolated influenza viruses. Arrows indicate known transmission routes between each species based on reported cases, phylogenetic analysis and experimental studies (adapted from Fereidouni et al., 2014).

No sequence of seal H4N5 virus is available in GenBank and only based on the HI assay of the virus and due to the high replication of the virus in avian hosts, it was concluded that the HA was antigenically and biologically similar to avian viruses (Hinshaw et al. 1984). In 1991, when two H4N6 influenza viruses were isolated from harbour seals, based on presence of H4 HA, it was concluded that they should originated from wild bird isolates (Callan et al. 1995).

In 1984, two whale influenza viruses, H13N2 and H13N9, were shown to be closely related to the H13 influenza viruses circulating among seagulls in the USA (Hinshaw et al. 1986). The seal H3N3 influenza isolate was first suspected to be related to seasonal human H3 influenza viruses, but later phylogenetic analysis demonstrated a close relationship to North American avian influenza viruses (Callan et al. 1995). The phylogenetic analyses of all eight segments of H3N8 seal influenza virus demonstrated also the closest relationship to waterfowl influenza viruses (Anthony et al. 2012).

At last, influenza viruses of H10N7 subtype, commonly found in migratory waterfowl, caused highest reported influenza-related mortality among harbour seals in 2014. Phylogenetic analyses showed that the HA of this virus is genetically closely related to Eurasian H10 viruses recently found in migratory ducks in Sweden, Denmark and Georgia (Zohari et al., 2014; Krog et al., 2015; Bodewes et al., 2015).

The 2010 influenza viruses isolated from free-ranging Northern elephant seals were the only influenza A viruses showed an homology to human and not avian viruses. Those viruses showed an homology greater than 99% to the pandemic influenza viruses (H1N1pdm09) isolated from humans in 2009 (Goldstein et al. 2013).

The evidence indicated that introduction of influenza viruses from wild birds and humans into marine mammal populations might have occurred on several occasions independently. High mortality of marine mammals, in several occasions, after transmission of avian influenza viruses to their population pose a great concern regarding the conservation of the population. Avian influenza viruses are known to be circulating at high prevalence in aquatic birds, and there is a continuous potential for transmission of viruses through direct or indirect contact of marine mammals with wild birds or their droppings at hauling-out sites or during feeding on the same food resources, although establishment of infection requires numerous factors (Fereidouni et al., 2014).

MARINE MAMMALS AND PUBLIC HEALTH

Circulation of avian influenza viruses in marine mammals may potentially have implications for public health, although the only reported evidence for transmission of influenza viruses from marine mammals to humans dates back to the 1979 outbreak of influenza in harbour seals with H7N7 subtype, which caused purulent keratoconjunctivitis in humans who handled the dead seals (Webster et al. 1981b). The virus raised concern as a zoonotic threat but it did not spread further. Affected people recovered without complications, and antibodies to the virus were not detectable in the serum samples of infected persons.

On the other hand, isolation of H1N1pdm09 influenza viruses from healthy seals in 2010 provided evidence for crossover transmission of influenza viruses from humans to marine mammals. The isolated viruses replicated in human epithelial respiratory cells not as efficient as human reference strains (Goldstein et al. 2013). This finding may indicating that these isolates adopted to elephant seal and lost potential to infect humans efficiently.

Genetic analyses of the isolated H3N8 viruses from harbour seals in 2011 indicated presence of mutations in the viral PB2 protein that increase transmissibility and virulence of influenza viruses in mammalian (Anthony et al. 2012). However, there is no evidence for transmission of those H3N8 viruses to other mammalian species, including humans.

Serological and virological data have shown that influenza B viruses could have persisted in the seal population for a long time without significant changes, a fact which may be considered as a risk for the re-introduction of the viruses to the human populations. However, at present no evidence is convincing enough to conclude that seals play a role as reservoir species for human influenza A or B viruses (Fereidouni et al., 2014).

CONCLUSION

In summary, interspecies transmission of influenza viruses is an important event in the evolution and ecology of these viruses. Influenza viruses are among the few zoonotic pathogens known to have caused infections in seals and whales. Infection of seals and other marine mammals with influenza A and B viruses have been reported on several occasions. Many constraints made it difficult to determine the real incidence of influenza virus infections in marine mammals. Nevertheless, data from several outbreaks and also serological results show that this is not a rare event, and probably the real prevalence has been underestimated (Fereidouni et al., 2014). Close genetic relatedness of influenza viruses isolated from marine mammals and wild birds suggests that wild birds are the main source of influenza virus infections of marine mammals (Figure 2). However, our knowledge is yet insufficient to make any general conclusion about transmission routes of the influenza viruses between wild birds and marine mammals or the role of them in ecological persistence of influenza viruses in the nature; therefore more detailed and validated monitoring as well as experimental studies are necessary to clarify the ecological and zoonotic importance of influenza viruses circulating in marine mammals.

REFERENCES

Anthony, SJ; St Leger, JA; Pugliares, K; Ip, HS; Chan, JM; Carpenter, ZW; et al. Emergence of fatal avian influenza in New England harbor seals. Mbio., 2012, 31 3(4), e00166-e001612.

Austin, FJ; Webster, RG. Evidence of ortho- and paramyxoviruses in fauna from Antarctica. J Wildl Dis., 1993, 29, 568-671.

Blanc, A; Ruchansky, D; Clara, M; Achaval, F; Le Bas, A; Arbiza, J. Serologic evidence of influenza A and B viruses in South American fur seals (Arctocephalus australis). J Wildl Dis., 2009, 45, 519-521.

Bodewes, R; Morick, D; de Mutsert, G; Osinga, N; Bestebroer, T; van der Vliet, S; et al. Recurring influenza B virus infections in seals. Emerg Infect Dis., 2013, 19, 511-512.

Bodewes, R; Bestebroer, TM; van der Vries, E; Verhagen, JH; Herfst, S; Koopmans, MP; et al. Avian Influenza A(H10N7) virus-associated mass deaths among harbor seals. Emerg Infect Dis., 2015, 21, 720-722.

de Boer, GF; Back, W; Osterhaus, AD. An ELISA for detection of antibodies against influenza A nucleoprotein in humans and various animal species. Arch Virol., 1990, 115, 47-61.

Callan, RJ; Early, G; Kida, H; Hinshaw, VS. The appearance of H3 influenza viruses in seals. J Gen Virol., 1995, 76, 199-203.

Calle, PP; Seagars, DJ; McClave, C; Senne, D; House, C; House, JA. Viral and bacterial serology of free-ranging Pacific walrus. J Wildl Dis., 2002, 38, 93-100.

Calle, PP; Seagars, DJ; McClave, C; Senne, D; House, C; House, JA. Viral and bacterial serology of six free-ranging bearded seals Erignathus barbatus. Dis Aquat Organ., 2008, 81, 77-80.

Fereidouni, S; Munoz, O; Von Dobschuetz, S; De Nardi, M. Influenza Virus Infection of Marine Mammals. Ecohealth., 2014, Sep 18.

Fernández, A; Esperón, F; Herraéz, P; de Los Monteros, AE; Clavel, C; Bernabé, A; et al. Morbillivirus and pilot whale deaths, Mediterranean Sea. Emerg Infect Dis., 2008, 14, 792-794.

Fouchier, RAM; Bestebroer, TM; Martina, BEE; Rimmelzwaan, GF; Osterhaus, ADME. Infection of grey seals and harbour seals with influenza B virus. International Congress Series., 2001, 1219, 225-231.

Fujii, K; Kakumoto, C; Kobayashi, M; Saito, S; Kariya, T; Watanabe, Y; et al. Serological evidence of influenza A virus infection in Kuril harbor seals (Phoca vitulina stejnegeri) of Hokkaido, Japan. J Vet Med Sci., 2007, 69, 259-263.

Geraci, JR; St Aubin, DJ; Barker, IK; Webster, RG; Hinshaw, VS; Bean, WJ; et al. Mass mortality of harbor seals: pneumonia associated with influenza A virus. Science., 1982, 215, 1129-1131.

Goldstein, T; Mena, I; Anthony, SJ; Medina, R; Robinson, PW; Greig, DJ; et al. Pandemic H1N1 influenza isolated from free-ranging Northern Elephant Seals in 2010 off the central California coast. PLoS One., 2013, 8(5), e62259.

Härkönen, T; Dietz, R; Reijnders, P; Teilmann, J; Harding, K; Hall, A; et al. The 1988 and 2002 phocine distemper virus epidemics in European harbour seals. Dis Aquat Organ., 2006, 68, 115-130.

Hinshaw, VS; Bean, WJ; Webster, RG; Rehg, JE; Fiorelli, P; Early, G; et al. Are seals frequently infected with avian influenza viruses? J Virol., 1984, 51, 863-865.

Hinshaw, VS; Bean, WJ; Geraci, JR; Fiorelli, P; Early, G; Webster, RG. Characterization of two influenza A viruses from a pilot whale. J Virol., 1986, 58, 655-656.

Kennedy, S; Kuiken, T; Jepson, PD; Deaville, R; Forsyth, M; Barrett, T; et al. Mass die-Off of Caspian seals caused by canine distemper virus. Emerg Infect Dis., 2000, 6, 637-639.

Kida, H; Brown, LE; Webster, RG. Biological activity of monoclonal antibodies to operationally defined antigenic regions on the hemagglutinin molecule of A/Seal/Massachusetts/1/80 (H7N7) influenza virus. Virol., 1982, 122, 38-47.

Kreutzer, M; Kreutzer, R; Siebert, U; Müller, G; Reijnders, P; Brasseur, S; et al. In search of virus carriers of the 1988 and 2002 phocine distemper virus outbreaks in European harbour seals. Arch Virol., 2008, 153, 187-192.

Krog, JS; Hansen, MS; Holm, E; Hjulsager, CK; Chriél, M; Pedersen, K; et al. Influenza A(H10N7) virus in dead harbor seals, Denmark. Emerg Infect Dis., 2015, 21, 684-687.

Lang, G; Gagnon, A; Geraci, JR. Isolation of an influenza A virus from seals. Arch Virol., 1982, 68, 189-195.

Lipscomb, TP; Schulman, FY; Moffett, D; Kennedy, S. Morbilliviral disease in Atlantic bottlenose dolphins (Tursiops truncatus) from the 1987-1988 epizootic. J Wildl Dis., 1994, 30, 567-571.

Lvov, DK; Zdanov, VM; Sazonov, AA; Braude, NA; Vladimirtceva, EA; Agafonova, LV; et al. Comparison of influenza viruses isolated from man and from whales. Bull World Health Organ., 1978, 56, 923-930.

Mamaev, LV; Visser, IK; Belikov, SI; Denikina, NN; Harder, T; Goatley, L; et al. Canine distemper virus in Lake Baikal seals (Phoca sibirica). Vet Rec., 1996, 138, 437-439.

Mandler, J; Gorman, OT; Ludwig, S; Schroeder, E; Fitch, WM; Webster, RG; et al. Derivation of the nucleoproteins (NP) of influenza A viruses isolated from marine mammals. Virol., 1990, 176, 255-261.

Naeve, CW; Webster, RG. Sequence of the hemagglutinin gene from influenza virus A/Seal/Mass/1/80. Virol., 1983, 129, 298-308.

Nielsen, O; Clavijo, A; Boughen, JA. Serologic evidence of influenza A infection in marine mammals of arctic Canada. J Wildl Dis., 2001, 37, 820-825.

Ohishi, K; Ninomiya, A; Kida, H; Park, CH; Maruyama, T; Arai, T; et al. Serological evidence of transmission of human influenza A and B viruses to Caspian seals (Phoca caspica). Microbiol Immunol., 2002, 46, 639-644.

Ohishi, K; Maruyama, T; Ninomiya, A; Kida, H; Zenitani, R; Bando, T; et al. Serologic investigation of influenza a virus infection in cetaceans from the western North Pacific and the Southern Oceans. Marine Mammal Science., 2006, 22, 214-221.

Osterhaus, AD; Rimmelzwaan, GF; Martina, BE; Bestebroer, TM; Fouchier, RA. Influenza B virus in seals. Science., 2000, 288, 1051-1053.

Stuen, S; Have, P; Osterhaus, AD; Arnemo, JM; Moustgaard, A. Serological investigation of virus infections in harp seals (Phoca groenlandica) and hooded seals (Cystophora cristata). Vet Rec., 1994, 134, 502-503.

Taubenberger, JK; Tsai, M; Krafft, AE; Lichy, JH; Reid, AH; Schulman, FY; et al. Two morbilliviruses implicated in bottlenose dolphin epizootics. Emerg Infect Dis., 1996, 2, 213-216.

Van Bressem, MF; Van Waerebeek, K; Raga, JA. A review of virus infections of cataceans and the potential impact of morbilliviruses, poxviruses and papillomaviruses on host population dynamics. Dis Aquat Organ., 1999, 38, 53-65.

Waltzek, TB; Cortes-Hinojosa, G; Wellehan, JF; Jr. Gray, GC. Marine mammal zoonoses: a review of disease manifestations. Zoonoses Public Health., 2012, 59, 521-535.

Webster, RG; Hinshaw, VS; Bean, WJ; Van Wyke, KL; Geraci, JR; St Aubin, DJ; et al. Characterisation of an influenza A virus from seals. Virol., 1981, 113, 712-724.

Webster, RG; Geraci, J; Petursson, G; Skirnisson, K. Conjunctivitis in human beings caused by influenza A virus of seals. New Engl J Med., 1981, 304, 911.

Webster, RG; Bean, WJ; Gorman, OT; Chambers, TM; Kawaoka, Y. Evolution and ecology of influenza A viruses. Microbiol Mol Biol Rev., 1992, 56, 152-179.

Zohari, S; Neimanis, A; Härkönen, T; Moraeus, C; Valarcher, JF. Avian influenza A(H10N7) virus involvement in mass mortality of harbour seals (Phoca vitulina) in Sweden, March through October 2014. Euro Surveill., 2014, 19, pii: 20967.

In: Advances in Marine Biology. Volume 1 ISBN: 978-1-53610-131-7
Editors: Adam Kovács and Patrik Nagy © 2016 Nova Science Publishers, Inc.

Chapter 2

MARINE SPONGE COMPOUNDS WITH ANTI-INFLAMMATORY ACTIVITY IN 2012–2016; AND THEIR MECHANISM OF ACTION

Maushmi Kumar*
SVKM'S NMIMS, Shobhaben Pratapbhai Patel School of Pharmacy and Technology Management, Mumbai, India

ABSTRACT

This chapter presents an overview of the development and study of marine sponge's bioactive compounds for anti-inflammatory activity for last four years i.e., 2012-2016. Marine pharmacology during 2009–2011 remained a global enterprise contributing to the preclinical pharmacology of 262 marine compounds which are in pharmaceutical pipeline. There is no updated review for the studies done on marine sponge's compounds showing anti-inflammatory activity after 2011. This chapter is a sincere effort to present a systematic review of the preclinical pharmacology of marine sponge and its associated microbes/symbionts compounds for potential activity in inflammation.

* Email: maushmiskumar@gmail.com.

Keywords: porifera, inflammation, inflammatory mediators, IL-1, IL-6, TNF-α, PGE2

1. Introduction

An inflammatory response is a natural defensive mechanism triggered by body tissues on being damaged by invading pathogens, physical stress, injury, persistent foreign bodies and by autoimmune reactions. Depending on the stimuli and the causative agents, it can be short or long lasting. Chronic inflammation involves the release of a number of mediators such as monocytes, macrophages, lymphocytes, plasma cells and fibroblasts that are not prominent in the acute response (Beg et al., 2011). The defence elements of body is found majorly in blood and during inflammation these cells and chemicals leave the blood and enter the tissue around the injured or infected site to protect and fight damage. Any type of infection caused by a bacterium, virus or fungus, commonly causes inflammation; while inflammation is the body's own response to it. It is part of many other common acute and chronic diseases. The therapeutic strategy for the treatment of patients with inflammation has been first, the relief of symptoms and the maintenance of function; and second the slowing or arrest of the tissue-damaging process. Nonsteroidal anti-inflammatory drugs (NSAIDs) relieves pain for significant periods, nonopioid analgesics are appropriate for the treatment of both acute and chronic inflammatory conditions. The glucocorticoids also have powerful anti-inflammatory effect but their toxicity limits their use mostly in cases of rheumatoid arthritis. The identified targets for inflammation are the pro-inflammatory mediators cyclo-oxygenase 1 and 2 (COX-1/-2), mitogen-activated protein kinases (MAPKs), janus protein tyrosine kinases (JAKs), nuclear transcription factor (NF-κB) and signal transducers and activators of transcription (STAT) which all directly or indirectly lead to the production of a vast number of pro-inflammatory cytokines and regulatory proteins such as IL-1/6, TNF-α, MIF, IFN-γ, MMPs (Kulkarni et al., 2006).

The role of natural products as therapy in inflammation has been recognized since ancient ages, and major progress has been done using combinatorial chemistry. Natural products with anti-inflammatory activity have been traditionally used for inflammatory conditions such as fevers, pain, migraine and arthritis (Yuan et al., 2006). Many such natural products and their derivatives with anti-inflammatory activity have been studied in detail like curcumin, parthenolide, resveratrol, cucurbitacins, 1, 8-cineole, salicylate, epigallocatechin-3-gallate, pseudopterosins, quercetin, lyprinol, bromelain, flavonoids, saponins and

Boswellia serrata gum resin. Recently there is also lot of interest in multi-target drug concept to use the huge array of natural products with privileged scaffolds for developing safe anti-inflammatory multi-target drugs (Koeberle et al., 2014). In investigation of extracts from a marine organisms- sea urchin, *Echinus esculentus* for anti-inflammatory activity, UPLC-ToF-MS and NMR analysis concluded the purified compound with a phosphocholine group with an unknown side chain. A bryozoan, *Eucratea loricata*'s organic extract revealed anti-inflammatory activity in NFκB, TNF-α and IL-1β screening (Do Thuy, 2012). There are many anti-inflammatory natural compounds reported from marine sponge, varying widely in both chemical structure and biological activity. The anti-inflammatory compounds from marine sponge are dominated by terpenoid compounds, especially sesterterpenes (Keyzers and Davis-Coleman, 2005).

In this chapter we review the development of marine sponge's anti-inflammatory compounds for last four years i.e., 2012-2016. A systematic review of the preclinical pharmacology of marine sponge and its associated microbes/ symbionts compounds for potential activity in inflammation is discussed.

2. Marine Sponge's Anti-Inflammatory Activity

Twenty four anti-inflammatory compounds were reported during the year 2008-11, which may be classified into diverse structural types: terpenes, non-terpenes, alkaloids and peptides. Marine sponges are the richest source of anti-inflammatory compounds among other marine invertebrates such as ascidian, soft coral, cnidarians and gorgonian. The anti-inflammatory sponge natural product is dominated by isoprenoid derived metabolites, especially sesterterpenes followed by nitrogenous compounds- alkaloids, peptides etc. (Keyzers and Davis-Coleman, 2005). New treatment approaches for acute and chronic inflammation is investigating novel anti-inflammatory molecules which can target Toll-like receptor (TLR) signaling pathways. Fung et al. (2014) worked on crude marine sponge extracts and identified compound Girolline which targeted TLR5 signalling. It inhibited signaling through both MyD88-dependent and independent TLRs (i.e., TLR2, 3, 4, 5, and 7) and reduced cytokine (IL-6 and IL-8) production in human peripheral blood mononuclear cells and macrophages. In other study, *Geodia cydonium* extract induced a reduction in VEGF (vascular endothelial growth factor) and proinflammatory cytokines (CCL2, CXCL8, CXCL10, IFN-γ, and TNF-α) levels in human breast cancer cell line MCF-7 cells, indicating an anti-inflammatory effect (Costantini et al., 2015). *Spongia officinalis* was investigated for in vivo anti-inflammatory activity using the carrageenan-induced paw edema model in rats. The methanol extract (25, 50 and 100 mg/kg) produced a significant reduction of the edema and the semi-purified fraction F3 from methanol extract exhibited significant activity at the dose of 50 mg/kg, at the third hour after carrageenan injection, with 72.85% reduction in paw volume (Dellai et al., 2012). The methanolic extract of the Red sea marine sponge *Xestospongia testudinaria* prevented carrageenan-induced acute local inflammation in rats. Methanolic extract (100 mg/kg) of red sea sponge *Xestospongia testudinaria* decreased % increase in paw weight measured at 1, 2, 3 and 4 h after carrageenan injection. The extract decreased paw malondialdehyde (MDA) and nitric oxide (NO) and decreased the inflammatory cytokines, tumor necrosis factor-α (TNF-α), interleukin-1 β (IL-1β) and IL-6 (El-Shitany et al., 2015). Phorbaketal A, a tricyclic sesterterpenoid was isolated from the marine sponge *Phorbas* sp. It significantly inhibited the production of nitric oxide (NO), by showing suppression in the expression of inducible NO synthase at both the mRNA and protein levels in LPS- induced RAW 264.7 cells. Further, it also reduced production of inflammatory cytokines such as tumor necrosis factor-alpha, interleukin (IL)-1beta, IL-6, and monocyte chemotactic protein-1 (Seo et al., 2015). The ethanol extract of the demosponge *Homaxinella* cf. *balfourensis* showed activity by reducing the release of IL-1β and PGE2 only at the highest, but not cytotoxic concentration of 250μg/mL significantly. The extract of *Isodictya erinacea* inhibited the release of IL-1β and PGE2, although not affecting LTB4 in a dose dependent manner. The extracts of the sponge *Isodictya toxophila* and *Mycale* (Oxymycale) *acerata*'s decreased the release of IL-1β, LTB4 and PGE2 release at the lower concentration of 50μg/mL, which was not cytotoxic (Moles et al., 2014). Solomonsterol A, a selective pregnane X receptor (PXR) agonist was isolated from the marine sponge *Theonella swinhoei*. It exerted anti-inflammatory activity along with attenuation of systemic inflammation and immune dysfunction in a mouse model of rheumatoid arthritis. It reduced the degree of joint damage by inhibiting the expression of pro-inflammatory mediators. (Mencarelli et al., 2014). In another study on the anti-inflammatory effect of compound 11-oxoaerothionin isolated from the Verongida sponge *Aplysina fistularis*, a concentration-dependent inhibition of iNOS (inducible nitric oxide synthase) protein, NO^-_2 (nitrite), proinflammatory cytokines and PGE2 (prostaglandin E2) production was observed, when RAW264.7 cells and primary macrophages were preincubated with the compound after stimulation with lipopolysaccharide (Medeiros et al., 2012).

Aeroplysinin-1 is a secondary metabolite isolated from marine sponge *Aplysina* genus with a wide spectrum of bio-activities. It is a chiral, optically active molecule. The enantiomer (-)-aeroplysinin-1 was isolated from *Ianthella ardis* and (+)-aeroplysinin-1 was the first brominated derivative from *Aplysina aerophoba* isolated by Fattorusso and Minale (1972). The dextrorotatory enantiomer has been investigated extensively. It has potent antibiotic effects on gram-positive bacteria and several dinoflagellate microalgae. In preclinical studies, (+)-aeroplysinin-1 have shown promising anti-inflammatory, anti-angiogenic and anti-tumor effects. Due to its versatility, it is of pharmaceutical interest for treatment of different pathologies (Garcia-Vilas et al., 2015). Another marine sponge *Neopetrosia* (order Haploscleridae, family Petrosiidae) whole crude extract and its isolated compounds revealed antimicrobial, anti-fouling, anti-HIV, cytotoxic, anti-tumor, anti-oxidant, anti-protozoal, anti-inflammatory activities. Total extracts of *N. proxima* (100 mg/Kg) and *N. rosariensis* (100 mg/kg) significantly inhibited the paw edema of rats 60% and 72% resp. Dichloromethane and methanol fractions of *N. proxima* reduced myeloperoxidase activity and dichloromethane fraction of *N. rosariensis* significantly inhibited nitric oxide (66%), prostaglandin E2 (30.5%) and tumor necrosis factor alpha production (72%) (Franco et al., 2012; Qaralleh 2016). The crude extract of sponge *Reniera sarai*, also displayed anti-inflammatory activities (Dhinakaran et al., 2014). Red sea sponges *Scalarispongia aqabaensis* and *Callyspongia siphonella* are poorly investigated genera for their compounds and biomedicinal value. Sterols scalaristerol (5α, 8α- dihydroxycholest-6-en-3β-ol) from *Scalarispongia aqabaensis*, and callysterol (ergosta-5, 11-dien-3β-ol) from *Callyspongia siphonella* were isolated for its anti-inflammatory activity in rat-hind paw edema. The results indicated that callysterol has a strong anti-inflammatory activity, which was comparable to cortisone, while compound scalaristerol showed moderate anti-inflammatory activity. Callysterol affected the release of superoxide anion and thromboxane B2 in a concentration-dependent manner. It showed close to 50% inhibition at approximately 10 μM (Youssef et al., 2010).

Table 1. Compounds isolated from marine sponge and fungus and their mechanism of action

Sr. no.	Compound	Sponge	Molecular mechanism	Reference
1.	Phorbaketal A 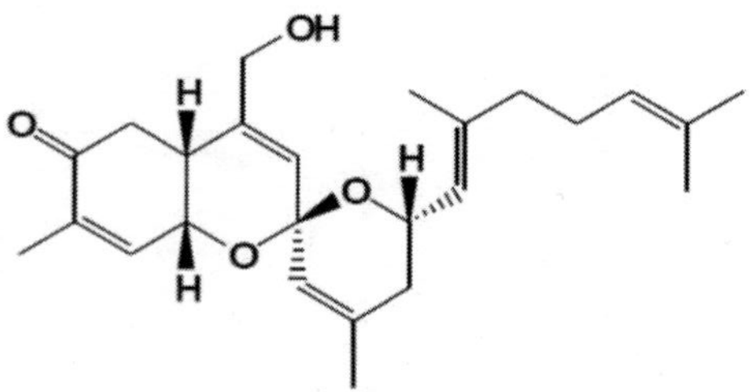	*Phorbas* sp.	Suppressed the expression of inducible NO synthase, reduced the LPS-induced production of inflammatory cytokines - tumor necrosis factor-alpha, interleukin (IL)-1beta, IL-6, and monocyte chemotactic protein-1.	Seo et al., 2015
2.	(+) Aeroplysinin- 1 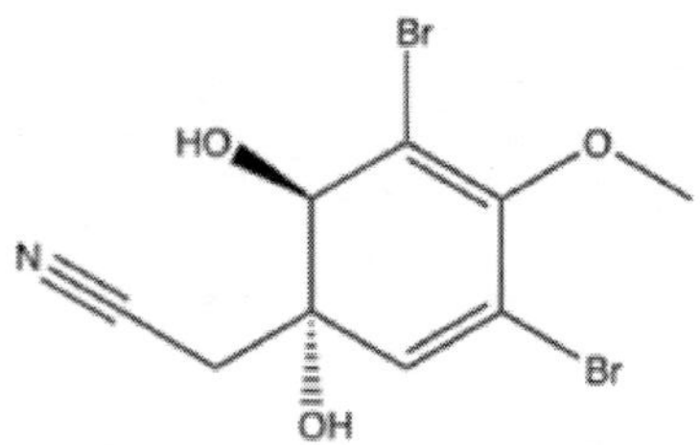	*Aplysina aerophoba*	Decreased the expression levels of mRNA and protein levels of MMP-2, MCP-1, TSP-1, COX-2, Il-1α and MMP-1.	Garcia-Vilas et al., 2015

Sr. no.	Compound	Sponge	Molecular mechanism	Reference
3.	Scalaristerol 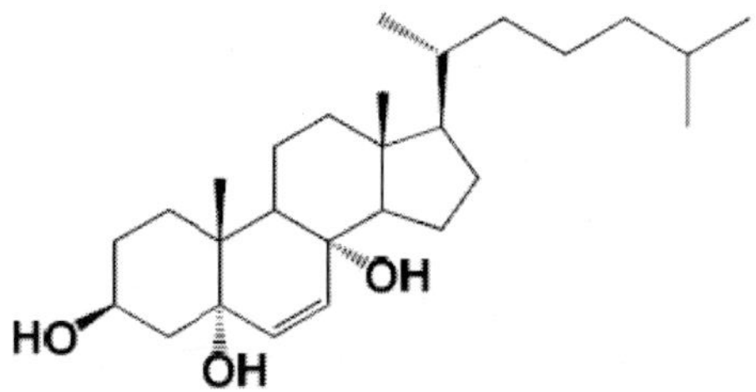	*Scalarispongia aqabaensis*	No effect on release of superoxide anion and thromboxane B2	Youssef at al., 2010
4.	Callysterol 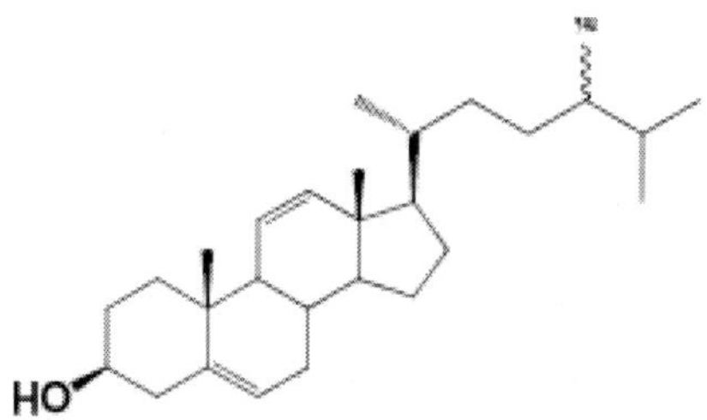	*Callyspongia siphonella*	Inhibition of release of superoxide anion and thromboxane B2	Youssef at al., 2010

Table 1. (Continued)

Sr. no.	Compound	Sponge	Molecular mechanism	Reference
5.	Solomonsterol A 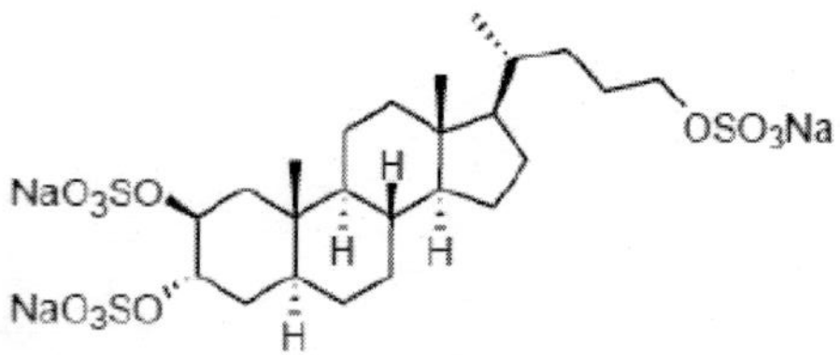	*Theonella swinhoei*	Reduces the expression of TNFα, IFNγ and IL-17 and chemokines-MIP1α and RANTES.	Mencarelli et al., 2014
6.	Dactyloquinone B 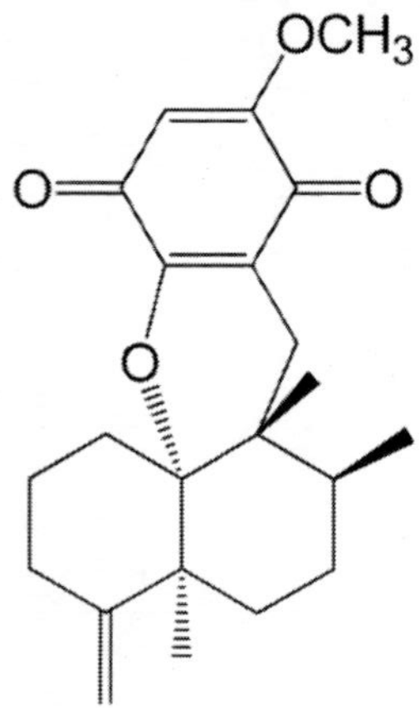	*Dactylospongi a elegans*	Mixture of both compounds decreased production of TNF-α, IL-1β, and IL-6 production. decreased the production of NO production, and iNOS and COX-2 expression	Lee et al., 2015

Sr. no.	Compound	Sponge	Molecular mechanism	Reference
7.	Cyclospongiaquinone-1 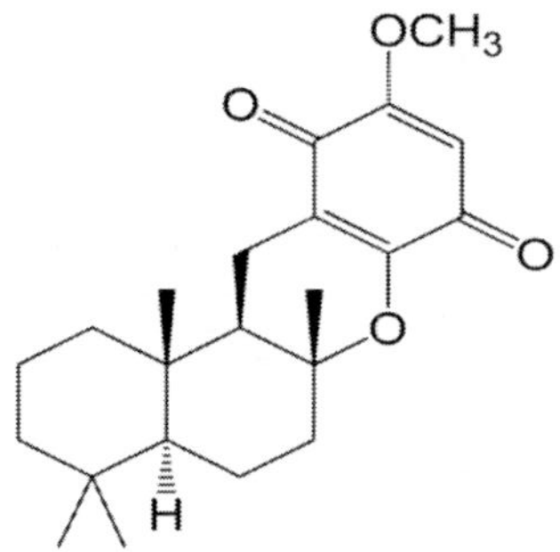 	*Stelospongia conulata*		Kazlauskas et al., 1978
8.	Mycoepoxydiene 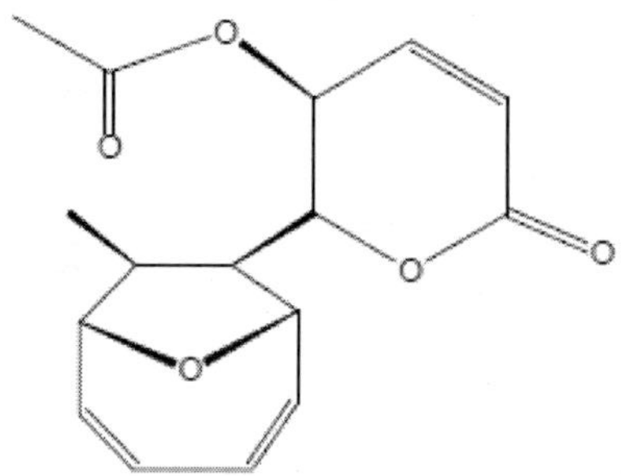	Marine fungus *Diaporthe* sp.	inhibition of NF-κB and MAPK pathway activation	Chen et al., 2012

3. Chemical Structures of Compounds Isolated and Their Mechanism of Action

A wide range of chemical compounds from various marine invertebrates have been investigated for their anti-inflammatory properties (Table 1). Most of them exhibit phospholipase A2 inhibitory activity. They also control nuclear factor-B activation and inflammatory gene expression (Alcaraz and Paya, 2006). Some sesquiterpenes were found to inhibit human neutrophil degranulation, superoxide generation, leukotriene B4 production (variabilin), TPA (12-o-tetradecanoylphorbol-13-acetate)-induced ear oedema (variabilin, bolinaquinone, topically) as well as carrageenan induced paw oedema (variabilin, petrosaspongiolide M and bolinaquinone, p.o.) in mice (Giannini et al., 2001; Haefner 2003).

4. Anti-Inflammatory Activities of Bacteria Associated with Marine Sponges

Anti-inflammatory activity of secondary metabolites produced by *Theonella* sp. was studied using RAW 264.7 macrophages. Inhibition of nitric oxide (NO) production in lipopolysaccharide stimulated RAW 264.7 cells is a well-established assay to screen anti-inflammatory activity. The inhibition level of NO released by RAW 264.7 was estimated from nitrate standard curve. Two isolates TM 1.8 and TM 1.9 possessed anti-inflammatory activities of 112.06% and 109.7% respectively for every 1ug/L sample. The result was compared in inhibiting NO production compared to positive control N-Monomethyl-L-arginine Monoacetate (L-NMMA) which showed an inhibitory activity at 87.41% for every 1ug/L sample (Radzi et al., 2015). New tanzawaic acid derivative tanzawaic acid Q, and four known analogues, tanzawaic acids A, C, D, and K were investigated from a marine-derived fungus *Penicillium steckii* 108YD142. These compounds significantly inhibited nitric oxide production and the new tanzawaic acid Q inhibited the lipopolysaccharide (LPS)-induced inducible nitric oxide synthase and cyclooxygenase-2 proteins and mRNA expressions in RAW 264.7 macrophages. The result of this study demonstrated that the new tanzawaic acid derivative inhibits LPS-induced inflammation (Shin et al., 2016).

5. Synthetic Compounds from the Marine Sponge Compounds Lead

Several inflammatory diseases most notably rheumatoid arthritis, is related to P2X7R activation, which has earned great attention and interests in the application of new treatments through the inhibition of this receptor (Toulme et al., 2010). The functional role of the P2X7R in the inflammatory event is well known as an important mediator in the expression and release of important cytokines and inflammatory mediators. Because of its significant role, it is recently considered as an important anti-inflammatory and pain target, and it has received lot of attention from pharmaceutical companies for evaluating clinical trials based on it (Soares-Bezerra et al., 2013).

Table 2. Synthetic compounds from the marine sponge compounds as a lead

	Target effects	Compounds	Reference
1.	P2X3R and P2X2/3R and.	AF-353 (Roche Pharmaceuticals) A-317491 (Abbott Laboratories)	Gever et al., 2006
2.	P2X7 receptors	Cyanoguanidines A-438079, A-740003 and A804598	Nelson et al., 2006
3.	Anti-inflammatory action in human neutrophils	Carteramine A	Kobayashi et al., 2007

Several new synthetic compounds have been developed for both the P2X3R and P2X2/3R and P2X7 receptors. AF-353 is the most notable synthetic compound developed by Roche Pharmaceuticals (Gever et al., 2006) and A-317491 developed by Abbott Laboratories for P2X3R (Gunosewoyo and Kassiou, 2010). Disubstituted tetrazoles, the cyanoguanidines A-438079, A-740003 and A804598 with action on hP2X7R (human)/mP2X7R (mouse) and rP2X7R (rat) are the candidates for a potential antagonist for P2X7R (Nelson et al., 2006). The compounds are listed in Table 2. TerraMarine Pharmaceuticals, New Zealand has already identified two promising anti-inflammatory compounds from marine organisms. The first one was patented in September 2005, and is now being improved for its effectiveness and usability. Further, they also plan to study the compound's action in the body and its pharmacokinetic properties. The study will help to determine the correct dosage of the compound and also its development into a usable drug.

TerraMarine is also in process to file the provisional patent for the second compound. Owning a patent will achieve TerraMarine the rights to use the chemical structure of the compound as an anti-inflammatory drug or any other for 16 years (Proffitt 2006).

CONCLUSION

Research work summarized in this chapter highlights the ongoing search and development in the field of marine sponge's bioactive compounds and their extraordinary potential as a source of novel bioactive compounds and drugs. In the meantime, much work is in progress with reference to the clinical trials and bringing the drug in the market. However, this ultimate step is dependent on the extent to which the pharmaceutical industry shows its preparedness to grasp such opportunities. The recent policy adopted by most governments worldwide is encouraging and aims at funding basic research, clinical and the manufacturing industry. There is need of continuous effort in preclinical and clinical studies of the marine anti-inflammatory compounds to be tested so that more and more marine drugs are launched. The continuous search for marine sponge leads, and its transformation into a drug with such financial support will ensue good results.

REFERENCES

Alcaraz, M. J. and Paya, M. (2006). Marine sponge metabolites for the control of inflammatory diseases. *Current Opinion in Investigational Drugs,* 7(11), 974-979.

Beg, S., Swain, S., Hasan, H., Barkat, M. A., Hussain, M. S. (2011). Systematic review of herbals as potential anti-inflammatory agents: Recent advances, current clinical status and future perspectives. *Pharmacognosy Reviews,* 5(10), 120–137.

Chen, Q., Chen, T., Li, W., Zhang, W., Zhu, J., Li, Y., Huang, Y., Shen, Y., Yu, C. (2012). Mycoepoxydiene inhibits lipopolysaccharide-induced inflammatory responses through the suppression of TRAF6 polyubiquitination. *PLoS ONE,* 7, e44890.

Costantini, S., Romano, G., Rusolo F., Capone, F., Guerriero, E., Colonna G., Ianora, A., Ciliberto, G., Costantini, M. (2015). Anti-inflammatory effects of a methanol extract from the marine sponge *Geodia cydonium* on the human breast cancer MCF-7 cell line. *Mediators of Inflammation*, 9, doi:10.1155/2015/204975.

Dellai, A., Deghrigue, M., Laroche-Clary, A., Masour, H.B., Chouchane, N., Robert, J., Bouraoui, A. (2012). Evaluation of antiproliferative and anti-inflammatory activities of methanol extract and its fractions from the Mediterranean sponge. *Cancer Cell International,* 12(18).

Dhinakaran D. I., Sivakumar T., Lipton A.P. (2014) Detection of Marine Derived Natural Products Synthesized in Marine Sponges Collected from Tuticorin, India. *European Journal of Biological Sciences*, 6 (1), 01-06.

Do, T. M-A. (2012). Isolation and Characterization of Anti-inflammatory Compounds from Marine Organisms: *Eucratea loricata* and *Echinus esculentus*. Master thesis for the degree Master of Pharmacy, University of TromsQ, UIT.

El-Shitany, N.A., Shaala, L.A., Abbas, A.T., Abdel-dayem, U.A., Azhar, E.I., Ali, S.S., Van Soest, R.W.M., Diaa, and Youssef, T.A. (2015) Evaluation of the anti-inflammatory, antioxidant and immunomodulatory effects of the organic extract of the Red Sea marine sponge *Xestospongia testudinaria* against Carrageenan induced rat paw inflammation. *PLoS ONE* 10(9), e0138917.

Fattorusso, E., Minale L.S.G. (1972) Aeroplysinin-1, an antibacterial bromo-compound from the sponge *Verongia aerophoba. Journal of Chemical Society,* 72(1), 16–18.

Franco, L.A., Macareno, J.L., Ocampo, Y.C., Pajaro, I.P., Gaitan, R. (2012). Marine sponges of the genus *Neopetrosia* with anti-inflammatory activity. *Latin American Journal of Pharmacy,* 31, 976-983.

Fung, S.Y., Sofiyev, V., Schneiderman, J., Hirschfeld, A. F., Victor, R.E., Woods, K., Piotrowski, J.S., Deshpande, R., Li, S.C., Voogd, N.J. de, Myers, C.L., Boone, C., Andersen, R. J., Turvey, S.E. (2014). Unbiased screening of marine sponge extracts for anti-inflammatory agents combined with chemical genomics identifies Girolline as an inhibitor of protein synthesis. *ACS Chemical Biology,* 9 (1), 247–257.

Garcia-Vilas, J.A., Martinez-Poveda, B., Quesada, A.R., Medina, M.A. (2016). Aeroplysinin-1, a sponge-derived multi-targeted bioactive marine drug. *Marine Drugs,* 14, 1.

Gever, J.R., Cockayne, D.A., Dillon, M.P., Burnstock, G., Ford, A.P. (2006). Pharmacology of P2X channels. *Pflugers Archiv- European Journal of Physiology,* 452, 513–537.

Giannini, C., Debitus, C., Lucas, R., Ubeda, A., Paya, M., Hooper, J.N., Dauria, M.V. (2001). New sesquiterpene derivatives from the sponge *Dysidea* species with a selective inhibitor profile against human phospholipase A2 and other leukocyte functions. *Journal of Natural products*, 64, 612–615.

Gunosewoyo, H., Kassiou, M. (2010). P2X purinergic receptor ligands: recently patented compounds. Expert Opinion on Therapeutic Patents, 20, 625–646. Haefner, B. (2003). Drugs from the deep: Marine natural products as drug candidates. *Drug Discovery Today,* 8, 536-544.

Haitham, Q. (2016). Chemical and bioactive diversities of marine sponge Neopetrosia. *Bangladesh Journal of Pharmacology,* 11, 433-452.

Kazlauskas, R., Murphy, P.T., Warren, R.G., Wells, R.J., Blount, J.F. (1978). New quinones from a dictyoceratid sponge. *Australian Journal of Chemistry,* 31(12), 2685 – 2697.

Keyzers, R. A. and Davies-Coleman M.T. (2005). Anti-inflammatory metabolites from marine sponges. *Chemical Society Reviews,* 34, 355-365.

Kobayashi, H., Kitamura, K., Nagai K., Nakaoa, Y., Fusetani, N., van Soest, R.W.M., Matsunaga, S. (2007). Carteramine A, an inhibitor of neutrophil chemotaxis, from the marine sponge *Stylissa carteri*. Tetrahedron Letters, 48(12), 2127–2129.

Koeberle, A. and Werz, O. (2014). Multi-target approach for natural products in inflammation. *Drug Discovery Today,* 19(12), 1871-1882.

Kulkarni, R.G., Achaiah, G., Sastry, G.N. (2006). Novel targets for anti-inflammatory and antiarthritic agents. *Current Pharmaceutical Design*, 12, 2437-54.

Medeiros A.I., Gandolfie, R.C., Secattob, A., Falcuccib, R.M., Facciolib, L.H., Hajduc, E., Peixinhod, S., Berlincke, R. G. S. (2012). 11-Oxoaerothionin isolated from the marine sponge *Aplysina fistularis* shows anti-inflammatory activity in LPS-stimulated macrophages. *Immunopharmacology and Immunotoxicology,* 34(6), 919-924.

Mencarelli, A., D'Amore, C., Renga B., Cipriani, S., Carino, A., Sepe, V., Perissutti, E., D'Auria, M.V., Zampella, A., Distrutti E. and Fiorucci, S. (2014). Solomonsterol A, a marine Pregnane-X-receptor agonist, attenuates inflammation and immune dysfunction in a mouse model of arthritis. *Marine Drugs,* 12, 36-53.

Menna, M., Imperatore, C., D'Aniello, F., Aiello, A. (2013). Meroterpenes from marine invertebrates: Structures, occurrence, and ecological implications. *Marine Drugs*, 11, 1602-1643.

Moles, J., Torrent, A., Alcaraz, M.J., Ruhi, R., Avila, C. (2014). Anti-inflammatory activity in selected Antarctic benthic organisms. *Frontiers in Marine Science*, 1(24), 1-5.

Nelson, D.W., Gregg, R.J., Kort, M.E., Perez-Medrano, A., Voight, E.A., Wang, Y., Grayson, G., Namovic, M.T., Donnelly-Roberts, D.L., Niforatos, W., Honore, P., Jarvis, M.F., Faltynek, C. R., Carroll, W. A. (2006). Structure-activity relationship studies on a series of novel, substituted 1-benzyl-5-phenyltetrazole P2X7 antagonists. *Journal of Medicinal Chemistry,* 49, 3659–3666.

Proffitt, F. (2006). New drugs from nature: the partnership. *Water and Atmosphere,* 14(3), 18-19. www.niwa.co.nz.

Radzia, S.A.M., Andriania, Y., Habsah M.A., Mohamada T.S.T., Saidina, J. (2015). In-vitro anti-inflammatory activities of extracts from bacteria associated with marine sponges: *Theonella* sp. 77(25), 165–169.

Shin, H.J., Pil G.B., Heo, S.J., Lee, H.S., Lee, J.S., Lee, Y-J., Lee, J., Won, H.S. (2016). Anti-inflammatory activity of tanzawaic acid derivatives from a marine-derived fungus *Penicillium steckii* 108YD142 *Marine Drugs,* 14(1), 14.

Soares-Bezerra, R.J., Calheiros, A.S., Ferreira, N.C. da Silva., Frutuoso, V. da Silva., Alves, L.A. (2013). Natural products as a source for new anti-inflammatory and analgesic compounds through the inhibition of purinergic P2x receptors. *Pharmaceuticals,* 6, 650-658.

Seo, Y-J., Lee, K.T., Rho, J.R., Choi, J.H. (2015). Phorbaketal A isolated from the marine sponge *Phorbas* sp., Exerts Its Anti-Inflammatory Effects via NF-κB inhibition and heme oxygenase-1 activation in lipopolysaccharide-stimulated macrophages. *Marine Drugs,* 13, 7005-7019.

Toulme, E., Tsuda, M., Khakh, B.S., Inoue, K. (2010). On the role of ATP-gated P2X receptors in acute, inflammatory and neuropathic pain. In: Translational Pain Research: From Mouse to Man; Kruger, L., Light, A.R., Eds; CRC Press: Boca Raton, FL, USA.

Youssefa, D.T.A., Ibrahim, A. K., Khalifa S.I., Mesbah, M. K., Mayer, A. M. S., van Soest R.W.M. (2010). New anti-inflammatory sterols from the Red sea sponges *Scalarispongia aqabaensis* and *Callyspongia siphonella. Natural Product Communications,* 5 (1), 27 – 31.

Yuan, G., Wahlqvist, M.L., He, G., Yang, M., Li, D. (2006). Natural products and anti-inflammatory activity. *Asia Pacific Journal of Clinical Nutrition,* 15(2), 143-152.

In: Advances in Marine Biology. Volume 1 ISBN: 978-1-53610-131-7
Editors: Adam Kovács and Patrik Nagy © 2016 Nova Science Publishers, Inc.

Chapter 3

HELMINTH DIVERSITY OF CETACEANS: AN UPDATE

Natalia Fraija-Fernández*, Mercedes Fernández, Juan A. Raga and Francisco J. Aznar
Marine Zoology Unit,
Cavanilles Institute of Biodiversity and Evolutionary Biology,
Science Park, University of Valencia, Valencia, Spain

ABSTRACT

Parasitism is an extraordinary life-strategy that largely influences that of most free-living organisms, including cetaceans. Parasites of cetaceans have a positive intrinsic and instrumental value, but also a potential pathogenic impact on host populations and, therefore, they should be considered in any biodiversity or conservation program. Yet, a starting point for understanding the relationships between cetaceans and their parasites is to have a detailed account on parasite diversity. The knowledge of the parasite fauna of cetaceans goes back to the time when commercial whaling was allowed and material for scientific purposes was available. Over the last decades, however, parasites have been obtained mainly from stranded or by-caught animals, and regular updates of host-parasite lists have been published until the 1990's. In this chapter we provide an updated review of the helminth fauna of cetaceans. The data

* E-mail address of corresponding author: natalia.fraija@uv.es.

here presented constitutes a baseline for future diversity surveys and a way to encourage researchers towards a greater awareness about the biological importance of the helminth fauna of cetaceans. Out of the 90 species currently recognized in the order Cetacea, 72 have hitherto been examined for helminths. The family Balaenopteridae among baleen whales and the family Delphinidae among toothed whales harbor the most diverse helminth fauna, which is partly accounted for by a higher sampling effort. In contrast, the helminth fauna of the beaked whales (family Ziphiidae) is the least known since only 9 out of the 22 species in the family have been examined for parasites. Currently, there are 174 helminth species reported in cetaceans, from which nematodes is the most speciose group (62 spp.), followed by digeneans (54 spp.), cestodes (38 spp.) and acanthocephalans (20 spp.). However, 20 (11.5%) of these species represent accidental infections that occur because cetaceans and other marine vertebrates (i.e., pinnipeds and marine birds) share common prey. The families Anisakidae, Pseudaliidae and Tetrameridae (Nematoda), Brachycladiidae, Notocotylidae, Heterophyidae and Brauninidae (Digenea), Diphyllobothriidae, Phyllobothriidae and Tetrabothriidae (Cestoda), and Polymorphidae (Acanthocephala) have a major species-representation in the helminth fauna of cetaceans.

PARASITISM IN CETACEANS

Parasitism is an extraordinary and widespread life-strategy because parasites are able to infect almost every living organism, including cetaceans. Previous studies have highlighted the importance of including parasites in biodiversity and conservation programs for their cetacean hosts (Aznar et al. 2010). Like any other organism, parasites have an intrinsic value and should be considered as an integral part of the Biosphere. Parasites also have an instrumental value, and have commonly been used as natural tags for their host populations. Studies concerning cetacean social behaviour (Balbuena and Raga 1994), diet (Blazekovic et al. 2015), distribution (Aznar et al. 1995), stock identity (Marigo et al. 2015) and health status (Aznar et al. 2005) have benefited from the use of parasites as biomarkers. For instance, Blazekovic et al. (2015) found a relationship between the prevalence of three nematode species, i.e., *Anisakis simplex* sensu stricto, *A. physeteris* and *A. pegreffii* and the population structure of four toothed whales in the eastern Adriatic. On the other hand, heavy parasite loads are often detected in apparently healthy cetaceans (Rogan et al. 1997), but some species may have a clear pathogenic impact on cetacean populations. For instance, digeneans of the genus *Nasitrema,* which usually occur in the pterygoid sinuses of some dolphin species, may cause brain injuries (Arbelo et al. 2013). Details about the negative impact of parasites on cetacean populations have been explored in previous studies (see e.g., Aznar et al. 2001a, Raga et al. 2009).

In any event, our understanding of the historical and present-day interactions between cetaceans and their parasites require a detailed and updated account of its biodiversity. Helminths, i.e., acanthocephalans, cestodes, nematodes and digeneans, constitute the most significant portion of the metazoan parasitic fauna of cetaceans (Figure 1). Records of helminths in cetaceans go back to the early 1900's, and efforts for updating data have regularly been made until the 1990's (e.g., Baylis 1932, Price 1932, Delyamure 1955, Tomilin 1967, Yamaguti 1971, Dailey and Brownell 1972, Gibson and Harris 1979, Raga 1994). In this chapter we provide the reader with an updated review of the helminth fauna occurring in cetaceans. This goal is justified for several reasons, including, 1) an increase, over the last two decades, of the sampling effort on poorly known cetaceans and unexplored areas; 2) significant taxonomic changes and re-arrangements of helminth taxa; and 3) an increased use of molecular markers, which have helped to unveil a significant portion of hidden biodiversity (i.e., cryptic species). In the next section, we outline major diversity patterns of the helminth fauna of cetaceans. A host-parasite list is then presented in Tables 1 and 2.

Helminths of Cetaceans

Currently, there are 174 helminth species reported from cetaceans, which can be grouped into 4 major taxa as follows: Acanthocephala (20 spp.), Cestoda (38 spp.), Nematoda (62 spp.) and Digenea (54 spp.) (Table 1). However, a total of 20 species from these taxa represent accidental infections, which occur because cetaceans and other marine vertebrates (e.g., marine birds or pinnipeds) share prey and parasite larvae are exchanged among hosts through the trophic webs. Accidental infections include 7 species of acanthocephalans, 4 of cestodes, 6 of nematodes and 3 of digeneans (Table 1). Most of these taxa are typical from pinnipeds and marine birds. Interestingly, several helminth species that reproduce in cetaceans also have congeneric taxa infecting pinnipeds and/or marine birds, i.e., *Corynosoma* spp. (Acanthocephala), *Tetrabothrius* spp., *Diplogonoporus* spp. and *Diphyllobothrium* spp. (Cestoda), and *Orthosplanchnus* spp. and *Ogmogaster* spp. (Digenea). Also, *Anisakis simplex* sensu lato, which is typical from cetaceans, has also been reported as adult in some pinnipeds (Brattey and

Stenson 1993). These observations illustrate the high potential for host-switching events between marine mammals and birds (see Fraija-Fernández et al. 2015 and references therein).

A total of 72 out of 90 species currently recognized within the order Cetacea have ever been examined for helminths. The family Balaenopteridae among baleen whales, and the family Delphinidae among toothed whales, have the most diverse helminth fauna, which is not particularly surprisingly because they concentrate the highest sampling effort (Table 2). The helminth fauna of beaked whales (family Ziphiidae), on the other hand, is very poorly known, and not a single parasitological datum exists for 9 out of the 22 species included in this family.

As noted above, the helminth fauna of cetaceans is, as a whole, highly specific. At a lower taxonomic scale, highly specific taxa have a major representation among digeneans (14 spp. reported in single cetacean species), followed by nematodes (13 spp.), cestodes (8 spp.) and acanthocephalans (3 spp.) (Table 1). Only 19 species (10.9% of the total) has been reported just once.

Among the species of Acanthocephala, only species from two genera of the family Polymorphidae, namely *Bolbosoma* and *Corynosoma*, occur in the intestine, and occasionally in the stomach, of cetaceans (Table 2). Three families of cestodes have species infecting cetaceans, namely Tetrabothriidae, Diphyllobothriidae, and Phyllobothriidae, whereas nematodes are represented by taxa of three families, i.e., Pseudaliidae, Anisakidae and Tetrameridae. Finally, the bulk of digeneans reported in cetaceans belong to four families: Brachycladiidae, Notocotylidae, Heterophyidae and Brauninidae (Table 2).

In the Tables 1 and 2 below, information for helminth species, locality and references for each record is organized according to each cetacean species. Cetacean taxonomy follows the lists of species and subspecies of marine mammals made by the Committee on Taxonomy (2014). Only taxonomically accepted parasite name species are included as they were verified with the WoRMS Editorial Board (2015) and the Global Biodiversity Information Facility database (www.gbif.org). Parasite taxa not identified to the species level are not included, except in the cases in which the genus of the named species has not been previously recorded in the cetacean species. References are organized in chronologically order and are identified by numbers in the reference list. The information here presented may constitute a baseline for diversity studies. We encourage researchers to carry out parasite surveys of poorly studied cetacean species and areas, paying special attention to cryptic diversity.

Table 1. List of the helminth species and their families found in cetaceans. Abbreviations: (1) Accidental infections: not adult specimens reported in cetaceans; the species typically infect other vertebrates (i.e., fish, birds, pinnipeds); (2) Specific to a single cetacean species; reported more than once; (3) single records

Acanthocephala		Cestoda			
Species	Family	Species	Family	Species	Family
Bolbosoma balaenae	Polymorphidae	*Diphyllobothrium fuhrmanni*	Diphyllobothriidae	*Strobilocephalus triangularis*	Tetrabothriidae
Bolbosoma brevicolle	Polymorphidae	*Diphyllobothrium lanceolatum*	Diphyllobothriidae	*Tentacularia* sp.[1]	Tentacularidae
Bolbosoma capitatum	Polymorphidae	*Diphyllobothrium latum*[b]	Diphyllobothriidae	*Tetrabothrius affinis*	Tetrabothriidae
Bolbosoma hamiltoni	Polymorphidae	*Diphyllobothrium orcini*[3]	Diphyllobothriidae	*Tetrabothrius arsenyevi*[2]	Tetrabothriidae
Bolbosoma nipponicum	Polymorphidae	*Diphyllobothrium phocarum*[1c]	Diphyllobothriidae	*Tetrabothrius curilensis*[2]	Tetrabothriidae
Bolbosoma tuberculata	Polymorphidae	*Diphyllobothrium polyrugosum*	Diphyllobothriidae	*Tetrabothrius diplosoma*[3]	Tetrabothriidae
Bolbosoma turbinella	Polymorphidae	*Diphyllobothrium stemmacephalum*	Diphyllobothriidae	*Tetrabothrius egregius*[2]	Tetrabothriidae
Bolbosoma vasculosum[a]	Polymorphidae	*Diplogonoporus balaenopterae*	Diphyllobothriidae	*Tetrabothrius forsteri*	Tetrabothriidae
Corynosoma alaskensis[2]	Polymorphidae	*Hexagonoporus calyptocephalus*[2]	Diphyllobothriidae	*Tetrabothrius hobergi*[3]	Tetrabothriidae
Corynosoma australe[1]	Polymorphidae	*Hexagonoporus giganticus*[2]	Diphyllobothriidae	*Tetrabothrius ruudi*	Tetrabothriidae
Corynosoma bullosum[1]	Polymorphidae	*Hexagonoporus physeteris*[3]	Diphyllobothriidae	*Tetrabothrius schaeferi*[2]	Tetrabothriidae
Corynosoma cameroni[a]	Polymorphidae	*Multiductus physeteris*[3]	Diphyllobothriidae	*Tetrabothrius wilsoni*	Tetrabothriidae
Corynosoma cetaceum	Polymorphidae	*Plicobothrium globicephalae*	Diphyllobothriidae	*Trigonocotyle globicephalae*	Tetrabothriidae
Corynosoma curilense[2]	Polymorphidae	*Monorygma grimaldii*[d]	Phyllobothriidae	*Trigonocotyle lintoni*	Tetrabothriidae
Corynosoma wegeneri[1]	Polymorphidae	*Phyllobothrium delphini*[d]	Phyllobothriidae	*Trigonocotyle prudhoei*[e]	Tetrabothriidae
Corynosoma reductum[1]	Polymorphidae	*Phyllobothrium physeteris*[3d]	Phyllobothriidae	*Trigonocotyle sexitesticulae*[3]	Tetrabothriidae
Corynosoma semerme[1]	Polymorphidae	*Scolex pleuronectis*[d]	Phyllobothriidae	*Trigonocotyle spasskyi*[2]	Tetrabothriidae
Corynosoma septentrionalis[2]	Polymorphidae	*Anophryocephalus* sp.[1]	Tetrabothriidae		
Corynosoma strumosum[1]	Polymorphidae	*Priapocephalus eschrichtii*[2]	Tetrabothriidae		
Corynosoma validum[1]	Polymorphidae	*Priapocephalus grandis*	Tetrabothriidae		
		Priapocephalus minor	Tetrabothriidae		

Table 1. (Continued)

Nematoda					
Species	Family	Species	Family	Species	Family
Anisakis berlandi	Anisakidae	*Procamallanus* sp.[1]	Pseudaliidae	*Stenurus auditivus*	Pseudaliidae
Anisakis brevispiculata	Anisakidae	*Halocercus brasiliensis*	Pseudaliidae	*Stenurus australis*[2]	Pseudaliidae
Anisakis insignis[2]	Anisakidae	*Halocercus dalli*[2]	Pseudaliidae	*Stenurus globicephalae*	Pseudaliidae
Anisakis nascettii	Anisakidae	*Halocercus delphini*	Pseudaliidae	*Stenurus minor*	Pseudaliidae
Anisakis paggiae	Anisakidae	*Halocercus hyperoodoni*[3]	Pseudaliidae	*Stenurus nanjingensis*[3]	Pseudaliidae
Anisakis pegreffii	Anisakidae	*Halocercus invaginatus*	Pseudaliidae	*Stenurus ovatus*	Pseudaliidae
Anisakis physeteris	Anisakidae	*Halocercus kirbyi*	Pseudaliidae	*Stenurus truei*[3]	Pseudaliidae
Anisakis simplex sensu lato[f]	Anisakidae	*Halocercus kleinenbergi*	Pseudaliidae	*Stenurus yamagutii*[3]	Pseudaliidae
Anisakis simplex sensu stricto	Anisakidae	*Halocercus lagenorhynchi*	Pseudaliidae	*Torynurus convolutus*	Pseudaliidae
Anisakis typica	Anisakidae	*Halocercus monoceris*	Pseudaliidae	*Torynurus dalli*[2]	Pseudaliidae
Anisakis ziphidarum	Anisakidae	*Halocercus pingi*	Pseudaliidae	*Hysterothylacium aduncum*[1]	Rhaphidascarididae
Contracaecum lobulatum[i]	Anisakidae	*Halocercus sunameri*[2]	Pseudaliidae	*Crassicauda anthonyi*	Tetrameridae
Contracaecum multipapillatum[1]	Anisakidae	*Halocercus taurica*	Pseudaliidae	*Crassicauda boopis*	Tetrameridae
Goezia sp.[1]	Anisakidae	*Pharurus alatus*	Pseudaliidae	*Crassicauda carbonelli*[2]	Tetrameridae
Phocascaris sp.[1]	Anisakidae	*Pharurus asiaeorientali*[2]	Pseudaliidae	*Crassicauda costata*[3]	Tetrameridae
Pseudoterranova ceticola	Anisakidae	*Pharurus pallasii*[2]	Pseudaliidae	*Crassicauda crassicauda*	Tetrameridae
Pseudoterranova decipiens[1]	Anisakidae	*Pharurus sunameri*[2]	Pseudaliidae	*Crassicauda delamureana*[3]	Tetrameridae
Pseudoterranova kogiae[2]	Anisakidae	*Pseudalius inflexus*	Pseudaliidae	*Crassicauda giliakiana*	Tetrameridae
Terranova sp.[3g]	Ascarididae	*Skrjabinalius cryptocephalus*	Pseudaliidae	*Crassicauda grampicola*	Tetrameridae
Porrocaecum sp.[g]	Monhysteridae	*Skrjabinalius guevarai*	Pseudaliidae	*Crassicauda magna*[2]	Tetrameridae
Odontobius ceti	Camallanidae	*Stenurus arctomarinus*[2]	Pseudaliidae	*Placentonema gigantissimum*[2]	Tetrameridae

Trematoda					
Species	Family	Species	Family	Species	Family
Brachycladium atlanticum	Brachycladiidae	*Nasitrema gondo*	Brachycladiidae	*Synthesium seymouri*[2]	Brachycladiidae
Brachycladium curilense	Brachycladiidae	*Nasitrema lanceolatum*	Brachycladiidae	*Synthesium subtile*	Brachycladiidae
Brachycladium delphini	Brachycladiidae	*Nasitrema spathulatum*[2]	Brachycladiidae	*Synthesium tursionis*	Brachycladiidae
Brachycladium goliath	Brachycladiidae	*Nasitrema stenosoma*	Brachycladiidae	*Braunina cordiformis*	Brauninidae
Brachycladium nipponicum	Brachycladiidae	*Nasitrema sunameri*[2]	Brachycladiidae	*Galactosomum erinaceum*[h]	Heterophyidae
Brachycladium pacificum	Brachycladiidae	*Odhneriella arctica*[2]	Brachycladiidae	*Pholeter gastrophilus*	Heterophyidae
Brachycladium palliatum	Brachycladiidae	*Odhneriella elongata*	Brachycladiidae	*Ogmogaster antarctica*	Notocotylidae
Brachycladium parvulum[3]	Brachycladiidae	*Orthosplanchnus arcticus*[2]	Brachycladiidae	*Ogmogaster grandis*	Notocotylidae
Brachycladium petrowi	Brachycladiidae	*Orthosplanchnus pygmaeus*[2]	Brachycladiidae	*Ogmogaster pentalineata*[2]	Notocotylidae
Campula folium[2]	Brachycladiidae	*Orthosplanchnus sudarikovi*[2]	Brachycladiidae	*Ogmogaster plicata*	Notocotylidae
Campula gondo	Brachycladiidae	*Oschmarinella albamarina*	Brachycladiidae	*Ogmogaster trilineata*[2]	Notocotylidae
Campula oblonga	Brachycladiidae	*Oschmarinella laevicaecum*	Brachycladiidae	*Amphimerus lancea*[b]	Opisthorchiidae
Cetitrema foliforme[2]	Brachycladiidae	*Oschmarinella macrorchis*[2]	Brachycladiidae	*Delphinicola tenuis*	Opisthorchiidae
Cetitrema meadi[3]	Brachycladiidae	*Oschmarinella rochebruni*	Brachycladiidae	*Opisthorchis tenuicollis*[b]	Opisthorchiidae
Hunterotrema caballeroi[2]	Brachycladiidae	*Oschmarinella sobolevi*	Brachycladiidae		
Hunterotrema macrosoma[3]	Brachycladiidae	*Synthesium delamurei*	Brachycladiidae		
Nasitrema attenuatum	Brachycladiidae	*Synthesium elongatum*	Brachycladiidae		
Nasitrema dalli	Brachycladiidae	*Synthesium mironovi*	Brachycladiidae		
Nasitrema delphini	Brachycladiidae	*Synthesium nipponicum*	Brachycladiidae		
Nasitrema globicephalae	Brachycladiidae	*Synthesium pontoporiae*[2]	Brachycladiidae		

a. Described only in cetaceans as immature individuals; adults unknown.

b. Probably accidental.

c. Synonym: *Pyramicocephalus phocarum.*

d. Collective name for larval forms of species of tetraphyllideans whose adult stage infects elasmobrachs; cetaceans act as intermediate hosts for some species.

e. Unclear taxonomical status (see Hoberg, 1990).

f. Name used for a species complex.

g. Synonym of *Pseudoterranova* (?).

h. Listed as "indetermined species" by Dailey et al. (2002). Likely accidental.

i. Species name not included in the taxonomic database searched.

Table 2. List of helminth taxa reported in cetaceans. Abbreviations: AO, Atlantic Ocean; PO, Pacific Ocean; IO, Indian Ocean; Arc, Arctic; Ant, Antarctic; MS, Mediterranean Sea; BlS, Black Sea; BaS, Baltic Sea; RS, Red Sea; AR, Amazon River; YR, Yangtze River; GR, Ganges River; SA, Region of South Africa; (§) Geographic location not specified by the authors; (*) See Table 1 for specific considerations

Acanthocephala	Locality	Ref.	Cestoda	Locality	Ref.	Nematoda	Locality	Ref.	Trematoda	Locality	Ref.
Mysticeti: Family Balaenidae *Balaena mysticetus* Linnaeus, 1758. Bowhead whale											
Bolbosoma balaenae	AO	[31, 76, 306]	*Monorygma grimaldii**	§	[17]	*Anisakis simplex* s.l.*	§	[137]	*Brachycladium goliath*	AO, PO	[31, 76, 314, 306, 272, 229]
			*Phyllobothrium delphini**	AO	[31, 76, 263, 84, 229]	*Odontobius ceti*	§	[274]	*Ogmogaster antarctica*	Arc	[258]
			*Phyllobothrium physeteris**	§	[31]	*Crassicauda crassicauda*	AO	[76, 306, 270]	*Ogmogaster plicata*	AO, PO	[285, 306, 124, 229]
			*Scolex pleuronectis**	AO	[306]						
Eubalaena australis (Desmoulins, 1822). Southern right whale											
-	-	-	*Priapocephalus grandis*	Ant, SA	[31]	*Crassicauda costata**	SA	[131]	-	-	-
			*Monorygma grimaldii**	§	[17]	*Odontobius ceti*	§	[31]			
			*Phyllobothrium delphini**	§	[17]						
Eubalaena glacialis (Müller, 1776). North Atlantic right whale											
Bolbosoma balaenae	§	[306]	*Diplogonoporus balaenopterae*	§	[306]	*Odontobius ceti*	PO	[262, 274]	*Ogmogaster antarctica*	§	[306]
Bolbosoma brevicolle	AO	[76]	*Monorygma grimaldii**	§	[17]						
Bolbosoma nipponicum	PO	[262, 306]	*Scolex pleuronectis**	§	[306]						
Bolbosoma turbinella	AO, PO	[84, 270, 306]	*Priapocephalus grandis*	§	[31, 76, 84, 306]						
			Tetrabothrius affinis	§	[306]						
			Tetrabothrius ruudi	PO	[77, 306]						

Acanthocephala	Locality	Ref.	Cestoda	Locality	Ref.	Nematoda	Locality	Ref.	Trematoda	Locality	Ref.
Family Balaenopteridae											
Balaenoptera acutorostrata Lacépède, 1804. Common minke whale											
Bolbosoma balaenae	AO	[31, 76, 306, 115]	*Diplogonoporus balaenopterae*	PO	[140, 115]	*Anisakis physeteris*	PO	[137, 209]	*Brachycladium goliath*	AO, PO, Ant	[42, 206, 31, 219, 76, 314, 306, 272, 229, 44]
Bolbosoma brevicolle	AO	[31, 76, 306, 84]	*Monorygma grimaldii**	§	[17]	*Anisakis simplex* s.l.*	AO	[31, 76, 73, 306, 170, 116]	*Ogmogaster antarctica*	Arc	[239]
Bolbosoma nipponicum	PO	[307, 76, 262, 313, 306]	*Phyllobothrium delphini**	§	[17]	*Anisakis simplex* s.s.	AO	[181, 180]	*Ogmogaster plicata*	AO, PO	[31, 219, 74, 76, 84, 306]
			Tetrabothrius sp.	Ant	[70]	*Pseudoterranova decipiens**	§	[76, 306]			
						Odontobius ceti	§	[274]			
						*Hysterothylacium aduncum**	Arc	[211]			
						Crassicauda crassicauda	AO	[76, 306, 271]			
Balaenoptera edeni Anderson, 1879. Bryde's whale											
Bolbosoma capitatum	AO	[166]	-	-	-	*Anisakis simplex* s.l.*	PO	[306, 115, 201]	*Brachycladium petrowi*	PO	[268, 306]
Bolbosoma tuberculata	IO	[266]				*Crassicauda crassicauda*	AO	[306]	*Cetitrema foliforme**	IO	[267, 314, 306]
Bolbosoma turbinella	PO	[201]							*Ogmogaster grandis*	IO	[306]

Table 2. (Continued)

Acanthocephala	Locality	Ref.	Cestoda	Locality	Ref.	Nematoda	Locality	Ref.	Trematoda	Locality	Ref.
						Balaenoptera borealis Lesson, 1828. Sei whale					
Bolbosoma balaenae	AO	[31, 76, 306]	*Diphyllobothrium* sp.	PO	[171]	*Anisakis simplex* s.l.*	AO, PO, Ant	[31, 172, 73, 171, 306, 115, 70, 201, 157]	*Brachycladium goliath*	AO, PO, Ant	[42, 206, 31, 76, 172, 289, 314, 171, 306, 272, 70, 200]
Bolbosoma brevicolle	AO	[31, 76, 306, 84]	*Diplogonoporus balaenopterae*	AO, Arc, Ant	[31, 76, 177, 312, 238, 306, 85, 229, 70]	*Crassicauda boopis*	AO, Ant	[14]	*Ogmogaster antarctica*	AO, PO, Arc	[239, 306, 200, 157, 109]
Bolbosoma nipponicum	PO	[76, 313, 306]	*Monorygma grimaldii**	§	[17]	*Crassicauda crassicauda*	AO, Ant	[31, 76, 162, 306, 271, 70, 200]	*Ogmogaster grandis*	Ant	[265, 306]
						*Crassicauda delamureana**	Ant	[131]			
Bolbosoma tuberculata	Ant	[266, 306]	*Scolex pleuronectis**	Ant	[306]	*Odontobius ceti*	PO	[245, 274]	*Ogmogaster plicata*	AO, PO, Arc	[31, 219, 74, 76, 172, 239, 314, 171, 306, 115]
Bolbosoma turbinella	AO, PO, Ant, SA	[30, 122, 31, 110, 76, 172, 171, 306, 270, 115, 70, 200, 201]	*Priapocephalus grandis*	AO, Ant	[31, 76, 177, 84, 306, 115, 229, 70]						

Acanthocephala	Locality	Ref.	Cestoda	Locality	Ref.	Nematoda	Locality	Ref.	Trematoda	Locality	Ref.
			Priapocephalus minor	AO	[23, 31, 76, 306, 229]						
			Tetrabothrius affinis	AO, Ant, SA	[31, 76, 306, 115, 229, 70, 157]						
			*Tetrabothrius arsenyevi**	Ant	[76, 312, 306]						
			Tetrabothrius wilsoni	AO, PO, Ant	[156, 31, 177, 115, 63, 229, 69]						
Balaenoptera musculus (Linnaeus, 1758). Blue whale											
Bolbosoma balaenae	AO, PO	[31, 76, 306]	*Diplogonoporus balaenopterae*	AO	[177, 238, 306, 115, 189]	*Anisakis simplex* s.l.*	AO, PO, IO	[31, 189, 76, 73, 306, 115]	*Brachycladium goliath*	AO	[306]
Bolbosoma brevicolle	AO, SA	[31, 76, 96, 306, 115]	*Monorygma grimaldii**	§	[17]	*Pseudoterranova decipiens**	§	[76]	*Ogmogaster antarctica*	Arc	[76, 239, 314, 306]
Bolbosoma hamiltoni	AO	[30, 31, 76, 313, 306, 115]	*Phyllobothrium delphini**	§	[17]	*Crassicauda boopis*	AO	[113, 115, 189]	*Ogmogaster grandis*	AO	[265, 306]
Bolbosoma nipponicum	PO	[78, 306]	*Priapocephalus grandis*	AO, PO, SA	[31, 76, 177, 220, 306, 115, 229]	*Crassicauda crassicauda*	AO, PO	[31, 76, 162, 306, 271, 113, 115]	*Ogmogaster plicata*	AO	[31, 219, 76]
Bolbosoma turbinella	AO, PO	[31, 76, 270, 306]	*Tetrabothrius affinis*	AO, PO	[31, 76, 220, 177, 229, 306, 115]	*Odontobius ceti*	PO	[31, 262]			
			*Tetrabothrius schaeferi**	AO	[189, 177, 306, 115]						
			Tetrabothrius wilsoni	AO	[177, 115]						

Table 2. (Continued)

Acanthocephala	Locality	Ref.	Cestoda	Locality	Ref.		Locality	Ref.	Trematoda	Locality	Ref.
Megaptera novaeangliae (Borowski, 1781). Humpback whale											
Bolbosoma balaenae	Ant	[243, 31, 76, 306]	*Diplogonoporus balaenopterae*	PO, Arc	[245, 238, 171, 306]	*Anisakis simplex* s.l.*	§	[73]	*Ogmogaster plicata*	AO	[115]
Bolbosoma nipponicum	PO	[78, 306]	*Monorygma grimaldii**	§	[17]	*Odontobius ceti*	§	[31, 274]			
Bolbosoma turbinella	AO, PO	[31, 76, 270, 306, 115]	*Phyllobothrium delphini**	§	[306]	*Pharurus alatus*	AO	[229]			
						Crassicauda boopis	PO	[31, 76, 162, 270, 306, 115]			
						Crassicauda crassicauda	AO	[76, 285, 306, 271]			
Balaenoptera physalus (Linnaeus, 1758). Fin whale											
Bolbosoma balaenae	AO	[76, 306, 115]	*Diplogonoporus balaenopterae*	Arc	[76, 177, 238, 306, 115]	*Anisakis simplex* s.l.*	AO, PO, Ant	[305, 285, 73, 115, 170, 201]	*Brachycladium goliath*	AO, PO, Ant	[76, 172, 314, 171, 306, 272]
Bolbosoma brevicolle	Ant, SA	[31, 76, 84, 306, 115]	*Monorygma grimaldii**	Ant	[269, 306]	*Odontobius ceti*	§	[31, 274]	*Brachycladium petrowi*	AO	[306]
Bolbosoma hamiltoni	Ant	[31, 76, 54, 313, 306, 115]	*Phyllobothrium delphini**	AO, PO	[172, 285, 171, 306, 170]	*Crassicauda boopis*	AO	[76, 113, 306, 115, 149, 116]	*Ogmogaster antarctica*	AO, PO, Arc, MS, SA	[76, 143, 239, 314, 171, 306, 115, 229, 167, 200, 201]
Bolbosoma nipponicum	PO	[76, 313, 306]	*Priapocephalus grandis*	AO, Ant, SA	[76, 177, 306, 115, 229]	*Crassicauda crassicauda*	AO, Ant, SA	[31, 76, 162, 306, 113, 271, 115, 200]	*Ogmogaster grandis*	Ant	[265, 306]

Acanthocephala	Locality	Ref.	Cestoda	Locality	Ref.	Nematoda	Locality	Ref.	Trematoda	Locality	Ref.
Bolbosoma turbinella	AO, PO	[76, 270, 306]	*Priapocephalus minor*	AO	[76, 306, 229]				*Ogmogaster plicata*	AO, PO, Arc	[31, 74, 76, 172, 262, 239, 314, 171, 306, 229]
			Tetrabothrius affinis	Ant	[161, 76, 306, 115]				*Ogmogaster trilineata**	PO	[242, 314, 171]
			*Tetrabothrius egregius**	Ant	[276, 306]						
			Tetrabothrius ruudi	Ant	[31, 76, 177, 306, 115]						
Family Neobalaenidae *Caperea marginata* (Gray, 1846). Pygmy right whale											
-	-	-	-	-	-	*Anisakis pegreffii*	SA	[181]	-	-	-
Family Eschrichtiidae *Eschrichtius robustus* (Lilljeborg, 1861). Gray whale											
Bolbosoma balaenae	PO	[66]	*Monorygma grimaldii**	§	[17]	*Anisakis simplex* s.l.*	PO	[246, 66]	*Brachycladium goliath*	PO	[246]
*Corynosoma semerme**	Arc	[246, 306]	*Phyllobothrium delphini**	§	[17]				*Ogmogaster antarctica*	PO	[246, 65, 171, 66]
*Corynosoma septentrionalis**	Arc	[254, 246, 306]	*Priapocephalus eschrichtii**	Arc	[202, 246, 306]				*Ogmogaster pentalineata**	PO, Arc	[239, 242, 314, 246, 306, 171, 115, 66]
*Corynosoma strumosum**	Arc	[246, 306]							*Ogmogaster plicata*	Arc	[246, 306]
*Corynosoma validum**	Arc	[246, 306]							*Orthosplanchnus pygmaeus**	PO	[316, 246, 306]

Table 2. (Continued)

Acanthocephala	Locality	Ref.	Cestoda	Locality	Ref.	Nematoda	Locality	Ref.	Trematoda	Locality	Ref.
Odontoceti: Family Delphinidae											
Cephalorhynchus commersonii (Lacépède, 1804). Commerson's dolphin											
-	-	-	*Monorygma grimaldii**	§	[17]	*Anisakis simplex* s.l.*	AO	[115, 36]	*Braunina cordiformis*	AO	[115, 36]
			*Phyllobothrium delphini**	§	[17]	*Halocercus brasiliensis*	AO	[115, 190]	*Pholeter gastrophilus*	AO	[36]
			Strobilocephalus triangularis	AO	[36]				*Synthesium delamurei*	AO	[35]
Cephalorhynchus eutropia (Gray, 1846). Chilean dolphin											
Corynosoma cetaceum	PO	[287, 20]	-	-	-	*Anisakis* sp.	PO	[287]	*Braunina cordiformis*	PO	[105, 287]
						Pseudoterranova sp.	PO	[287]	*Nasitrema globicephalae*	PO	[43]
									Synthesium tursionis	PO	[287]
Cephalorhynchus heavisidii (Gray, 1828). Haviside's dolphin											
-	-	-	*Monorygma grimaldii**	AO	[17]	*Anisakis pegreffii*	SA	[181]	-	-	-
			*Phyllobothrium delphini**	AO	[17]						
Cephalorhynchus hectori (Van Beneden, 1881). Hector's dolphin											
Corynosoma sp.	PO	[188]	*Monorygma grimaldii**	PO	[17]	*Contracaecum* sp.	PO	[188]	*Campula* sp.	§	[128]
			*Phyllobothrium delphini**	PO	[188, 17]	*Phocascaris* sp.*	PO	[188]	*Braunina cordiformis*	PO	[188]

Acanthocephala	Locality	Ref.	Cestoda	Locality	Ref.	Nematoda	Locality	Ref.	Trematoda	Locality	Ref.
Delphinus capensis Gray, 1828. Long-beaked common dolphin											
-	-	-	*Monorygma grimaldii**	§	[17]	*Anisakis simplex* s.l.*	AO, PO	[194, 147]	-	-	-
			*Phyllobothrium delphini**	§	[17]						
Delphinus delphis Linnaeus, 1758. Short-beaked common dolphin											
*Bolbosoma vasculosum**	MS, BlS	[295, 76, 77, 306]	*Diphyllobothrium lanceolatum*	§	[218]	*Anisakis pegreffii*	AO, MS	[181]	*Amphimerus lancea**	AO	[166]
Corynosoma cetaceum	AO, IO, BlS	[134, 76, 77, 313, 254, 306, 34]	*Diphyllobothrium stemmacephalum*	BlS	[31, 77, 38]	*Anisakis simplex* s.l.*	AO, PO, BlS, SA	[76, 77, 73, 306, 71, 115, 1, 116, 38, 21, 34]	*Brachycladium atlanticum*	MS	[178]
*Corynosoma strumosum**	§	[218]	*Monorygma grimaldii**	AO, PO, BlS	[23, 76, 77, 306, 281, 62, 71, 115, 67, 229, 1, 116]	*Anisakis simplex* s.s.	AO	[181]	*Brachycladium delphini*	AO, MS, BlS	[31, 219, 76, 77, 127, 314, 306, 229]
			*Phyllobothrium delphini**	AO, PO, BlS	[31, 119, 76, 77, 263, 303, 171, 306, 281, 62, 71, 115, 67, 229, 1, 116]	*Anisakis typica*	AO, PO, BlS	[31, 76, 77, 73, 306, 71]	*Brachycladium palliatum*	AO, BlS	[31, 219, 76, 77, 314, 306, 229, 116, 38]
			*Scolex pleuronectis**	PO, MS	[222, 71]	*Crassicauda crassicauda*	AO	[306]	*Braunina cordiformis*	AO, BlS	[76, 77, 276, 289, 314, 306, 229, 34, 109]
			Strobilocephalus triangularis	PO	[67]	*Goezia* sp.*	§	[237]	*Campula oblonga*	AO	[5, 116]

Table 2. (Continued)

Acanthocephala	Locality	Ref.	Cestoda	Locality	Ref.	Nematoda	Locality	Ref.	Trematoda	Locality	Ref.
			*Tetrabothrius diplosoma**	AO	[306]	*Halocercus delphini*	AO, PO, MS, BIS	[31, 94, 310, 76, 77, 283, 306, 115, 1, 116, 278]	*Galactosomum erinaceum**	AO, BIS	[31, 76, 77, 306, 229]
			Tetrabothrius forsteri	PO, MS, BIS	[31, 119, 76, 77, 306, 67, 222]	*Halocercus invaginatus*	§	[218, 13]	*Nasitrema delphini*	PO	[204, 314, 171, 12, 71, 62, 278]
						Halocercus kleinenbergi	BIS, SA	[76, 77, 306, 38]	*Nasitrema globicephalae*	PO	[62, 71, 67]
						Halocercus lagenorhynchi	AO, IO	[190, 286]	*Nasitrema lanceolatum*	PO	[62]
						*Pseudoterranova decipiens**	§	[218]	*Nasitrema stenosoma*	PO	[71]
						Skrjabinalius cryptocephalus	PO, BIS	[76, 77, 278, 38]	*Oschmarinella rochebruni*	AO, PO, BIS	[31, 219, 76, 77, 314, 247, 306, 62, 71, 67, 229, 116, 34]
						Skrjabinalius guevarai	AO, MS	[227, 190]	*Pholeter gastrophilus*	AO, BIS	[306, 92, 213, 229, 116, 38, 34]
						Stenurus minor	§	[218]	*Synthesium delamurei*	MS	[222]
									Synthesium tursionis	PO	[280]

Acanthocephala	Locality	Ref.	Cestoda	Locality	Ref.	Nematoda	Locality	Ref.	Trematoda	Locality	Ref.
Feresa attenuata Gray, 1874. Pygmy killer whale											
*Bolbosoma vasculosum**	PO	[194]	*Monorygma grimaldii**	PO	[194]	*Anisakis simplex* s.l.*	PO	[319]	*Nasitrema lanceolatum*	PO	[106]
			*Phyllobothrium delphini**	§	[17]	*Anisakis typica*	PO	[319, 184]			
			*Scolex pleuronectis**	PO	[106]	*Terranova* sp.*	PO	[194]			
			Tetrabothrius forsteri	PO	[106]	*Halocercus brasiliensis*	PO	[120]			
			*Trigonocotyle sexitesticulae**	PO	[126, 194]	*Stenurus globicephalae*	PO	[106, 194]			
Globicephala macrorhynchus Gray, 1846. Short-finned pilot whale											
Bolbosoma capitatum	AO	[194, 55]	*Anophryocephalus* sp.*	§	[226]	*Anisakis simplex* s.l.*	AO, SA	[138, 137, 115, 55]	*Brachycladium nipponicum*	PO	[308, 314, 306]
			Diphyllobothrium sp.	AO	[47]	*Anisakis simplex* s.s.	AO	[48]	*Campula gondo*	PO	[308, 314, 306]
			*Monorygma grimaldii**	AO	[194, 55]	*Anisakis typica*	AO	[194, 181, 47, 48]	*Nasitrema attenuatum*	PO	[204]
			*Phyllobothrium delphini**	AO	[17, 47]	*Crassicauda* sp.	AO	[194]	*Nasitrema globicephalae*	AO, PO	[204, 314, 171, 12, 196]
			Plicobothrium globicephalae	§	[315]	*Halocercus kleinenbergi*	AO	[47]	*Nasitrema gondo*	PO	[309, 314, 306]
			Trigonocotyle globicephalae	§	[312]	*Stenurus globicephalae*	AO	[94, 306, 196, 194, 55, 47]	*Nasitrema lanceolatum*	PO	[204, 314, 171]
			Trigonocotyle lintoni	PO	[306]	*Stenurus minor*	AO	[196]			
						Torynurus convolutus	AO, PO	[94]			

Table 2. (Continued)

Acanthocephala	Locality	Ref.	Cestoda	Locality	Ref.	Nematoda	Locality	Ref.	Trematoda	Locality	Ref.
					Globicephala melas (Traill, 1809). Long-finned pilot whale						
Bolbosoma capitatum	AO, MS	[31, 76, 313, 58, 306, 115, 95, 27, 232, 28, 25]	*Diphyllobothrium polyrugosum*	AO	[27]	*Anisakis berlandi*	SA	[181, 180]	*Campula gondo*	PO	[76]
			Diphyllobothrium stemmacephalum	AO	[26]	*Anisakis pegreffii*	PO	[180]	*Nasitrema globicephalae*	PO	[63, 99]
			*Monorygma grimaldii**	AO, MS	[76, 306, 229, 232]	*Anisakis simplex* s.l.*	AO	[88, 232, 182, 1, 116]	*Orthosplanchnus arcticus**	AO	[59, 229, 232]
			*Phyllobothrium delphini**	AO, MS	[119, 76, 263, 303, 306, 288, 115, 229, 232, 1]	*Anisakis simplex* s.s.	AO, SA	[115, 182, 181, 180]	*Pholeter gastrophilus*	AO, MS	[237, 229, 232]
			Plicobothrium globicephalae	AO	[241, 229, 232]	*Anisakis typica*	PO	[31, 76, 73, 306, 184]	*Synthesium delamurei*	AO, MS	[230, 229, 232, 28]
			*Scolex pleuronectis**	AO	[59]	*Crassicauda carbonelli**	MS	[231, 232]	*Synthesium subtile*	AO	[29, 232, 28, 35]
			Trigonocotyle globicephalae	AO, MS, SA	[23, 306, 115, 27, 28]	*Stenurus globicephalae*	AO, MS	[31, 76, 59, 306, 13, 115, 237, 232, 1, 116]			
			Trigonocotyle lintoni	AO, MS	[76, 59, 306, 229]	*Torynurus convolutus*	AO, PO	[31, 76, 306, 13]			

Acanthocephala	Locality	Ref.	Cestoda	Locality	Ref.	Nematoda	Locality	Ref.	Trematoda	Locality	Ref.
Grampus griseus (G. Cuvier, 1812). Risso's dolphin											
-	-	-	*Monorygma grimaldii**	PO	[69, 237, 229, 58]	*Anisakis pegreffii*	MS	[39]	*Brachycladium palliatum*	AO	[318, 229]
			*Phyllobothrium delphini**	AO	[31, 76, 263, 303, 318, 306, 229, 1, 194, 58]	*Anisakis physeteris*	PO	[147]	*Nasitrema gondo*	PO	[197]
			*Scolex pleuronectis**	MS	[98, 100, 9, 17]	*Anisakis simplex* s.l.*	PO, SA	[115, 147]	*Synthesium delamurei*	MS	[98]
			Tetrabothrius forsteri	MS	[98]	*Crassicauda grampicola*	AO, PO, MS	[135, 76, 162, 306, 235, 225, 197, 279]	*Synthesium tursionis*	AO	[318]
			Trigonocotyle globicephalae	MS	[98]	*Pseudoterranova ceticola*	PO	[147]	*Pholeter gastrophilus*	AO, MS	[318, 229, 58, 98]
						Stenurus globicephalae	AO, MS	[13, 237, 98]			
						Stenurus minor	AO	[285, 306, 190]			
Lagenodelphis hosei Fraser, 1956. Fraser's dolphin											
Bolbosoma sp.	PO	[284]	*Monorygma grimaldii**	AO, IO	[186, 294, 194, 55, 47]	*Anisakis pegreffii*	AO	[48]	*Campula* sp.	IO	[186]
			*Phyllobothrium delphini**	AO	[281, 294, 229, 194, 55, 47]	*Anisakis simplex* s.l.*	AO, IO	[186, 294]	*Nasitrema* sp.	PO	[278]
			Strobilocephalus triangularis	AO	[223, 294, 229]	*Anisakis typica*	AO	[48]			

Table 2. (Continued)

Acanthocephala	Locality	Ref.	Cestoda	Locality	Ref.	Nematoda	Locality	Ref.	Trematoda	Locality	Ref.
			Tetrabothrius forsteri	AO	[194]	*Stenurus ovatus*	PO	[186, 278]			
Lagenorhynchus acutus (Gray, 1828). Atlantic white-sided dolphin											
Bolbosoma sp.	AO	[37]	*Diphyllobothrium stemmacephalum*	AO	[208]	*Anisakis simplex* s.l.*	AO	[137, 115, 237, 116]	*Oschmarinella laevicaecum*	AO	[37]
			*Monorygma grimaldii**	AO	[23, 31, 76, 115, 229, 248, 116]	*Anisakis typica*	AO	[137]	*Pholeter gastrophilus*	AO	[37, 229, 248]
			*Phyllobothrium delphini**	AO	[23, 306, 248, 116]	*Pseudalius inflexus*	AO	[115, 190]			
			Strobilocephalus triangularis	AO	[23, 31, 76, 306, 229]	*Stenurus globicephalae*	AO	[306, 13, 37, 248]			
			Tetrabothrius forsteri	AO	[37, 229]	*Torynurus convolutus*	AO	[115]			
						Crassicauda grampicola	AO	[112]			
Lagenorhynchus albirostris (Gray, 1846). White-beaked dolphin											
-	-	-	*Monorygma grimaldii**	§	[17]	*Anisakis simplex* s.l.*	AO	[31, 76, 73, 306, 116]	*Pholeter gastrophilus*	AO	[116]
			*Phyllobothrium delphini**	§	[17]	*Anisakis simplex* s.s.	AO	[182]			
			Plicobothrium globicephalae	AO	[170]	*Halocercus lagenorhynchi*	AO	[31, 94, 310, 76, 306]			
						Pseudoterranova sp.	§	[15]			
Lagenorhynchus australis (Peale, 1848). Peale's dolphin											
-	-	-	*Trigonocotyle prudhoei**	AO	[177, 115]	-	-	-	-	-	-
Lagenorhynchus cruciger (Quoy and Gaimard, 1824). Hourglass dolphin											
-	-	-	*Phyllobothrium delphini**	AO	[103]	*Anisakis simplex* s.l.*	AO	[76, 306, 103]	*Oschmarinella* sp.	AO	[103]
			*Trigonocotyle prudhoei**	Ant	[306]	*Anisakis typica*	AO	[76, 306]	*Synthesium subtile*	AO	[35]

Acanthocephala	Locality	Ref.	Cestoda	Locality	Ref.	Nematoda	Locality	Ref.	Trematoda	Locality	Ref.
Lagenorhynchus obliquidens Gill, 1865. Pacific white-sided dolphin											
-	-	-	*Diphyllobothrium fuhrmanni*	PO	[26]	*Anisakis simplex* s.l.*	PO	[62, 71]	*Brachycladium* sp.	PO	[247]
			*Monorygma grimaldii**	PO	[67, 71]	*Crassicauda* sp.	PO	[67]	*Nasitrema globicephalae*	PO	[204, 171, 12, 62, 71, 67]
			*Phyllobothrium delphini**	PO	[136, 171, 71, 67]						
			*Scolex pleuronectis**	PO	[71]						
			Strobilocephalus triangularis	PO	[71]						
			Tetrabothrius sp.	PO	[71]						
Lagenorhynchus obscurus (Gray, 1828). Dusky dolphin											
*Corynosoma australe**	AO	[72]	*Monorygma grimaldii**	§	[17]	*Anisakis simplex* s.l.*	AO	[134, 72, 21]	*Braunina cordiformis*	AO, PO	[297, 72]
			*Phyllobothrium delphini**	AO, PO, SA	[115, 297, 201, 160]	*Anisakis typica*	SA	[31, 73, 115]	*Nasitrema* sp.	PO	[297]
			*Scolex pleuronectis**	PO	[177, 297]				*Pholeter gastrophilus*	AO, PO	[297, 72]
			*Trigonocotyle prudhoei**	AO	[177, 115]				*Synthesium subtile*	AO	[35]
Lissodelphis borealis (Peale, 1848). Northern right-whale dolphin											
-	-	-	*Monorygma grimaldii**	PO	[71]	*Anisakis berlandi*	PO	[181]	*Nasitrema globicephalae*	PO	[204, 314, 171, 12, 62, 71]
			*Phyllobothrium delphini**	§	[17]	*Anisakis simplex* s.l.*	PO	[62, 71]			
			*Scolex pleuronectis**	PO	[199]						
Lissodelphis peronii (Lacépède, 1804). Southern right-whale dolphin											
-	-	-	*Strobilocephalus triangularis*	PO	[201]	-	-	-	*Delphinicola tenuis*	PO	[201]
			Tetrabothrius forsteri	PO	[201]						

Table 2. (Continued)

Acanthocephala	Locality	Ref.	Cestoda	Locality	Ref.	Nematoda	Locality	Ref.	Trematoda	Locality	Ref.
Orcaella brevirostris (Owen in Gray, 1866). Irrawaddy dolphin											
-	-	-	-	-	-	-	-	-	*Amphimerus lancea**	PO, IO	[31, 219, 76, 306]
Orcinus orca (Linnaeus, 1758). Killer whale											
Bolbosoma nipponicum	PO	[262, 306]	*Diphyllobothrium fuhrmanni*	PO	[257]	*Anisakis simplex* s.s.	PO	[181]	*Oschmarinella albamarina*	AO	[114, 116]
Bolbosoma capitatum	AO, PO	[262, 306, 114, 116]	*Diphyllobothrium orcini**	PO	[123]	*Anisakis simplex* s.l.*	AO, PO	[76, 73, 171, 306, 114, 116]	*Synthesium subtile*	AO, PO, Arc	[261, 314, 306, 292, 35]
			Diphyllobothrium polyrugosum	AO	[83, 306, 115, 229, 114, 116]						
			*Monorygma grimaldii**	AO	[89, 229]						
			*Phyllobothrium delphini**	AO	[115, 229]						
			*Trigonocotyle spasskyi**	PO	[76, 306]						
Peponocephala electra (Gray, 1846). Melon-headed whale											
Bolbosoma capitatum	AO	[55]	*Diphyllobothrium* sp.	AO	[194]	*Anisakis physeteris*	PO	[147]	*Nasitrema gondo*	PO	[195]
			*Monorygma grimaldii**	AO	[55, 47]	*Anisakis simplex* s.l.*	PO	[147]			
			*Phyllobothrium delphini**	AO	[47]	*Anisakis typica*	AO	[129, 184]			
			Strobilocephalus triangularis	AO	[47]	*Halocercus* sp.	PO	[278]			
						Stenurus globicephalae	AO, PO	[194, 190, 278, 47]			

Acanthocephala	Locality	Ref.	Cestoda	Locality	Ref.	Nematoda	Locality	Ref.	Trematoda	Locality	Ref.
Pseudorca crassidens (Owen, 1846). False killer whale											
Bolbosoma capitatum	AO, IO	[31, 76, 313, 306, 115, 97, 11]	-	-	-	*Anisakis berlandi*	PO	[182]	*Nasitrema attenuatum*	PO	[204, 314, 171]
						Anisakis simplex s.l.*	AO	[76, 73, 318, 306, 115, 182, 11]	*Nasitrema globicephalae*	PO	[204, 314, 171, 12]
						Anisakis simplex s.s.	PO	[182]	*Nasitrema gondo*	PO	[198]
						Anisakis typica	AO, PO	[318, 209]	*Odhneriella elongata*	AO, PO	[229]
						Stenurus auditivus	AO	[318]	*Synthesium elongatum*	AO	[318]
						Stenurus globicephalae	AO	[190, 320]			
Sotalia fluviatilis (Gervais and Deville in Gervais, 1853). Tucuxi											
-	-	-	*Monorygma grimaldii**	§	[17]	*Anisakis simplex* s.l.*	AO	[21]	*Amphimerus lancea**	AO, AR	[31, 219, 76, 289, 115]
			*Phyllobothrium delphini**	§	[17]	*Anisakis typica*	AO	[253, 184, 192, 317]	*Braunina cordiformis*	AO	[253, 192]
						Halocercus brasiliensis	AO	[115, 253, 190, 192]	*Nasitrema attenuatum*	AO	[115]
Sotalia guianensis (Van Bénedén, 1864). Guiana dolphin											
-	-	-	-	-	-	*Anisakis typica*	AO	[130, 163]	*Braunina cordiformis*	AO	[163, 176, 93]
						Halocercus brasiliensis	AO, AO	[94, 310, 76, 306,192, 176, 47, 163, 120, 93]	*Nasitrema attenuatum*	AO	[95]
									Synthesium tursionis	AO	[176]
Sousa chinensis (Osbeck, 1765). Indo-Pacific humpback dolphin											
-	-	-	*Monorygma grimaldii**	PO	[17]	*Halocercus pingi*	PO	[212]	-	-	-
			*Phyllobothrium delphini**	PO	[17]						

Table 2. (Continued)

Acanthocephala	Locality	Ref.	Cestoda	Locality	Ref.	Nematoda	Locality	Ref.	Trematoda	Locality	Ref.
Stenella attenuata (Gray, 1846). Pantropical spotted dolphin											
Bolbosoma balaenae	PO	[68]	*Monorygma grimaldii**	PO	[68, 67, 207]	*Anisakis simplex* s.l.*	PO	[65, 68]	*Brachycladium pacificum*	PO, SA	[68, 115, 67]
*Bolbosoma vasculosum**	PO	[68]	*Phyllobothrium delphini**	PO	[68, 67]	*Anisakis typica*	AO	[181, 48]	*Braunina cordiformis*	PO	[76, 314, 306]
			*Scolex pleuronectis**	PO	[207]	*Halocercus delphini*	PO	[68]	*Nasitrema globicephalae*	PO	[67, 207]
			Strobilocephalus triangularis	PO	[68, 67]				*Nasitrema stenosoma*	PO	[204]
			Tetrabothrius forsteri	PO	[68, 67, 207]				*Oschmarinella rochebruni*	PO, SA	[68, 115, 67]
									Synthesium elongatum	SA	[115]
Stenella clymene (Gray, 1850). Clymene dolphin											
Bolbosoma sp.	AO	[47]	*Monorygma grimaldii**	AO	[17, 47]	*Anisakis* sp.	AO	[47]	-	-	-
			*Phyllobothrium delphini**	AO	[17, 47]	*Anisakis typica*	AO	[129]			
			*Scolex pleuronectis**	AO	[47]	*Halocercus brasiliensis*	AO	[47, 120]			
			Strobilocephalus triangularis	AO	[47]	*Skrjabinalius guevarai*	AO	[6]			
			Tetrabothrius forsteri	AO	[47]						
			Trigonocotyle sp.	AO	[47]						
Stenella frontalis (G. Cuvier, 1829). Atlantic spotted dolphin											
*Bolbosoma vasculosum**	AO	194	*Phyllobothrium delphini**	AO	[194, 47]	*Anisakis typica*	AO	[194]	*Brachycladium palliatum*	AO	[318]
			Strobilocephalus triangularis	AO	[47]	*Halocercus delphini*	AO	[318]	*Braunina cordiformis*	AO	[318]
									Pholeter gastrophilus	AO	[318]

Acanthocephala	Locality	Ref.	Cestoda	Locality	Ref.	Nematoda	Locality	Ref.	Trematoda	Locality	Ref.
						Stenella coeruleoalba (Meyen, 1833). Striped dolphin					
Bolbosoma capitatum	AO	[251]	*Monorygma grimaldii**	AO, PO, MS	[115, 67, 233, 229, 169, 1, 116, 58, 17, 47, 207]	*Anisakis pegreffii*	MS	[181, 180, 39]	*Brachycladium atlanticum*	AO, MS	[4, 99, 178]
Bolbosoma turbinella	AO	[251]	*Phyllobothrium delphini**	AO, PO, MS, SA	[92, 318, 115, 229, 169, 1, 116, 58, 17, 47, 207]	*Anisakis simplex* s.l.*	AO, PO, MS	[138, 73, 318, 169, 1, 116]	*Brachycladium delphini*	AO, MS	[127, 92, 229]
*Bolbosoma vasculosum**	MS	[233, 169, 8, 179]	*Scolex pleuronectis**	AO, PO, MS	[8, 251, 9, 17, 207, 159]	*Anisakis simplex* s.s.	AO	[181, 39]	*Brachycladium pacificum*	PO	[67]
			Strobilocephalus triangularis	AO, PO, MS	[68, 69, 233, 169, 229, 8, 207, 179]	*Anisakis typica*	AO, MS, SA	[73, 115, 3, 184, 48]	*Brachycladium palliatum*	AO, PO, MS	[233, 158, 229, 169, 207]
			Tetrabothrius forsteri	AO, PO, MS, SA	[115, 233, 229, 169, 8, 207, 179]	*Halocercus brasiliensis*	AO	[251]	*Nasitrema globicephalae*	PO	[67, 207]
			Trigonocotyle globicephalae	MS	[179]	*Halocercus delphini*	AO, MS	[299, 237]	*Oschmarinella rochebruni*	AO, MS	[224, 229, 169, 99, 116, 178, 109]
						Halocercus lagenorhynchi	AO, PO, SA	[252, 115, 116, 207]	*Pholeter gastrophilus*	AO, MS	[92, 233, 234, 229, 169, 168, 58, 159]
						Skrjabinalius guevarai	AO, MS	[233, 227, 169]			
						Stenurus minor	MS	[16]			
						Stenurus ovatus	MS	[169]			

Table 2. (Continued)

Acanthocephala	Locality	Ref.	Cestoda	Locality	Ref.	Nematoda	Locality	Ref.	Trematoda	Locality	Ref.
					Stenella longirostris (Gray, 1828). Spinner dolphin						
Bolbosoma balaenae	PO	[68]	*Diphyllobothrium fuhrmanni*	PO	[76, 306]	*Anisakis simplex* s.l.*	PO	[68]	*Brachycladium nipponicum*	PO	[308, 76, 306]
Bolbosoma hamiltoni	PO	[7]	*Diphyllobothrium stemmacephalum*	PO	[26]	*Anisakis typica*	AO, PO	[7, 48, 129]	*Brachycladium pacificum*	PO	[76, 68, 67, 7, 207]
*Bolbosoma vasculosum**	PO	[68]	*Monorygma grimaldii**	AO, PO	[68, 67, 55, 47]	*Halocercus brasiliensis*	AO	[47]	*Oschmarinella laevicaecum*	PO	[308, 76, 314, 306]
			*Phyllobothrium delphini**	AO, PO	[68, 67, 47]	*Halocercus delphini*	PO	[68]	*Oschmarinella rochebruni*	PO	[68, 67, 7]
			*Scolex pleuronectis**	AO, PO	[7, 47]				*Synthesium tursionis*	PO	[7]
			Strobilocephalus triangularis	AO, PO	[68, 67, 7, 47]						
			Tetrabothrius forsteri	AO, PO	[68, 67, 47]						
			Trigonocotyle sp.	AO	[47]						
					Steno bredanensis (G. Cuvier in Lesson, 1828). Rough-toothed dolphin						
Bolbosoma capitatum	AO, MS	[31, 76, 313, 306, 115]	*Phyllobothrium delphini**	AO	[55]	*Anisakis simplex* s.l.*	PO	[138, 73, 209, 147]	*Brachycladium palliatum*	AO	[107]
			Strobilocephalus triangularis	AO	[76, 306, 229]	*Anisakis simplex* s.s.	AO	[48]	*Nasitrema attenuatum*	AO	[95]
			Tetrabothrius forsteri	AO	[23, 31, 76, 306, 107]	*Anisakis typica*	AO	[107, 55, 48, 129]	*Nasitrema globicephalae*	AO	[95]
			*Trigonocotyle prudhoei**	AO	[177, 306, 115]				*Synthesium tursionis*	AO	[107]
									Braunina cordiformis	AO	[253]
									Pholeter gastrophilus	AO	[107]

Acanthocephala	Locality	Ref.	Cestoda	Locality	Ref.	Nematoda	Locality	Ref.	Trematoda	Locality	Ref.
Tursiops aduncus (Ehrenberg, 1833). Indo-Pacific bottlenose dolphin											
-	-	-	*Diphyllobothrium fuhrmanni*	RS	[142]	*Anisakis typica*	RS	[142]	*Brachycladium curilense*	RS	[142]
						Crassicauda sp.	IO	[151]	*Nasitrema attenuatum*	RS	[142]
						Halocercus lagenorhynchi	IO	[286]	*Pholeter gastrophilus*	RS	[142]
						Stenurus ovatus	IO	[286]			
Tursiops truncatus (Montagu, 1821). Common bottlenose dolphin											
Bolbosoma sp.	§	[40]	*Diphyllobothrium fuhrmanni*	PO	[257]	*Anisakis pegreffii*	MS, SA	[182, 39]	*Brachycladium atlanticum*	AO	[222]
Corynosoma australe[*]	AO	[250]	*Diphyllobothrium stemmacephalum*	BlS	[81, 86, 38]	*Anisakis simplex* s.l.[*]	AO, PO, BlS	[76, 115, 40, 1, 116, 38, 201, 250]	*Brachycladium delphini*	AO	[318]
Corynosoma cetaceum	AO, IO, BlS	[134, 76, 77, 313, 254, 306, 250]	*Monorygma grimaldii*[*]	AO, PO, MS, BlS	[31, 76, 77, 89, 306, 229, 298, 1, 116, 58, 207]	*Anisakis simplex* s.s.	AO	[48, 39]	*Brachycladium palliatum*	AO	[304, 318, 229]
			Phyllobothrium delphini[*]	AO, PO, MS, BlS	[31, 76, 77, 89, 263, 303, 306, 229, 298, 1, 116, 58, 55]	*Anisakis typica*	AO, PS, MS	[318, 73, 298, 184, 55, 48]	*Braunina cordiformis*	AO, PO, BlS	[76, 276, 255, 136, 314, 318, 171, 306, 118, 298, 253, 38, 250]
			Scolex pleuronectis[*]	AO	[17, 222]	*Contracaecum multipapillatum*[*]	AO	[48]	*Campula oblonga*	AO	[318]

Table 2. (Continued)

Acanthocephala	Locality	Ref.	Cestoda	Locality	Ref.	Nematoda	Locality	Ref.	Trematoda	Locality	Ref.
			Strobilocephalus triangularis	MS	[222]	*Crassicauda crassicauda*	MS, BIS	[31, 76, 77, 162, 306]	*Nasitrema attenuatum*	AO	[145, 115]
			Tetrabothrius forsteri	PO	[298, 222]	*Crassicauda grampicola*	BIS	[38]	*Nasitrema dalli*	SA	[115]
						Halocercus delphini	AO	[1, 190]	*Nasitrema delphini*	AO	[318]
						Halocercus lagenorhynchi	AO, PO, BIS	[135, 76, 77, 282, 304, 193, 318, 306, 115, 278]	*Nasitrema globicephalae*	PO	[298]
						Pharurus alatus	IO	[286]	*Pholeter gastrophilus*	AO, PO, MS, BIS	[304, 193, 92, 115, 234, 229, 298, 38, 222, 250, 109]
						Pseudoterranova decipiens[*]	AO	[250]	*Synthesium tursionis*	AO, PO, MS, BIS	[31, 219, 76, 77, 306, 229, 298, 102, 101, 99, 194, 280, 38, 174, 222, 250]
						Skrjabinalius cryptocephalus	PO, BIS	[40, 190, 278]			
						Skrjabinalius guevarai	MS	[111, 237, 190]			
						Stenurus minor	AO, PO, BIS	[190]			
						Stenurus ovatus	AO, PO, MS, BIS, IO	[31, 94, 75, 76, 77, 306, 115, 298, 278, 38, 286]			

Acanthocephala	Locality	Ref.	Cestoda	Locality	Ref.	Nematoda	Locality	Ref.	Trematoda	Locality	Ref.
						Family Iniidae *Inia geoffrensis* (Blainville, 1817). Amazon river dolphin					
-	-	-	-	-	-	*Anisakis insignis**	AR	[31, 76, 152, 115, 163, 317]	*Hunterotrema caballeroi**	AR	[187, 152, 304, 317]
									*Hunterotrema macrosoma**	AR	[60]
									Pholeter gastrophilus	AR	[317]
						Family Kogiidae *Kogia sima* (Owen, 1866). Dwarf sperm whale					
-	-	-	*Monorygma grimaldii**	§	[17]	*Anisakis brevispiculata*	AO	[55, 48, 221]	-	-	-
			*Phyllobothrium delphini**	AO, SA	[318, 115, 55, 47]	*Anisakis paggiae*	AO	[181, 55, 48, 221]			
						Anisakis physeteris	PO, SA	[115, 147]			
						Anisakis simplex s.l.*	AO, PO, SA	[115, 147, 48]			
						Anisakis typica	AO	[318, 48, 221]			
						Anisakis ziphidarum	AO	[221]			
						Pseudoterranova ceticola	AO, PO	[147, 55, 48]			

Table 2. (Continued)

Acanthocephala	Locality	Ref.	Cestoda	Locality	Ref.	Nematoda	Locality	Ref.	Trematoda	Locality	Ref.
Kogia breviceps (Blainville, 1838). Pygmy sperm whale											
*Bolbosoma vasculosum**	§	[215, 15]	*Monorygma grimaldii**	AO, IO	[23, 31, 115, 229, 1, 47]	*Anisakis brevispiculata*	AO, SA	[2, 181, 48]	-	-	-
			*Phyllobothrium delphini**	AO, MS∞, SA	[76, 263, 91, 303, 318, 306, 281, 115, 229, 1, 194, 55, 47]	*Anisakis paggiae*	AO, SA	[183, 48]			
			*Scolex pleuronectis**	AO	[9]	*Anisakis physeteris*	AO, PO, SA	[115, 194, 147, 48]			
						Anisakis simplex s.l.*	AO, PO, SA	[296, 73, 115, 1, 147, 48]			
						Anisakis simplex s.s.	AO	[48]			
						Anisakis typica	AO, SA	[318, 115, 129]			
						Pseudoterranova ceticola	AO, PO	[194, 147, 2, 48]			
						*Pseudoterranova kogiae**	SA	[76, 91, 306, 115]			
						*Crassicauda magna**	AO, PO	[76, 162, 91, 306, 1]			
∞Single record of the host species in the Mediterranean Sea											
Family Monodontidae. *Monodon monoceros* Linnaeus, 1758. Narwhal											
-	-	-	-	-	-	*Anisakis simplex* s.l.*	AO	[31, 76, 73, 306]	-	-	-
						*Pseudoterranova decipiens**	AO	[76, 306]			
						Halocercus monoceris	Arc	[302, 165]			
						Pharurus alatus	Arc	[31, 94, 76, 306, 13]			

Acanthocephala	Locality	Ref.	Cestoda	Locality	Ref.	Nematoda	Locality	Ref.	Trematoda	Locality	Ref.
					Delphinapterus leucas (Pallas, 1776). Beluga						
Bolbosoma sp.	AO	[191]	*Diphyllobothrium lanceolatum*	Arc	[82, 80, 306]	*Anisakis simplex* s.l.*	AO, Arc	[31, 164, 76, 73, 306, 115, 301, 191]	*Odhneriella arctica**	Arc	[82, 141, 80, 314, 306, 292, 301]
*Corynosoma cameroni**	AO, Arc	[295, 203, 306, 293]	*Diphyllobothrium phocarum**	Arc	[82, 306]	*Anisakis simplex* s.s.	AO	[182]	*Orthosplanchnus sudarikovi**	Arc	[290, 306, 292, 293]
*Corynosoma wegeneri**	Arc	[203, 84]	*Monorygma grimaldii**	§	[17]	*Crassicauda giliakiana*	PO	[76, 162, 80, 306]	*Oschmarinella albamarina*	Arc	[291, 314, 306, 292, 293]
*Corynosoma reductum**	AO	[295]	*Phyllobothrium delphini**	§	[17]	*Halocercus monoceris*	AO, Arc	[191, 148]	*Synthesium mironovi*	AO, PO	[219, 144, 76, 141, 285, 80, 314, 306]
*Corynosoma semerme**	PO, Arc	[203, 171, 293, 301]				*Halocercus taurica*	AO	[191, 148]	*Synthesium seymouri**	AO, PO, Arc	[31, 309, 76, 141, 80, 171, 306, 292, 293, 301, 191, 174]
*Corynosoma strumosum**	AO, PO, Arc	[164, 76, 203, 171, 306, 293, 301]				*Pharurus pallasii**	AO, PO, Arc	[76, 306, 13, 293, 301, 191]	*Synthesium subtile*	PO, Arc	[261, 306, 292]
						Porrocaecum sp.*	AO	[82, 256]			
						*Pseudoterranova decipiens**	Arc	[32, 80, 306]			
						*Stenurus arctomarinus**	AO, Arc	[306, 13, 293, 301, 191, 148]			
						Stenurus minor	AO	[80, 306]			

Table 2. (Continued)

Family Phocoenidae											
***Phocoena dioptrica* Lahille, 1912. Spectacled porpoise**											
*Corynosoma australe**	AO	[33]	*Scolex pleuronectis**	AO	[33, 205]	*Anisakis simplex* s.l.*	AO	[217, 33]	-	-	-
			Tetrabothrius sp.	AO	[205]						
Acanthocephala	Locality	Ref.	Cestoda	Locality	Ref.	Nematoda	Locality	Ref.	Trematoda	Locality	Ref.
Neophocaena phocaenoides (G. Cuvier, 1829). Indo-Pacific finless porpoise											
Corynosoma sp.	PO	[146]	*Diphyllobothrium fuhrmanni*	PO, IO	[306, 76, 133]	*Anisakis simplex* s.l.*	PO	[147]	*Campula folium**	PO	[210, 76, 314, 306, 133]
			*Monorygma grimaldii**	§	[17]	*Anisakis typica*	§	[133]	*Campula oblonga*	PO	[146, 133]
			*Phyllobothrium delphini**	§	[17]	*Halocercus pingi*	PO	[31, 94, 310, 76, 306, 146, 212, 133]	*Delphinicola tenuis*	PO	[146, 133]
						*Halocercus sunameri**	PO	[310, 306, 146, 133]	*Nasitrema spathulatum**	PO	[210, 309, 76, 314, 306, 146, 133]
						Halocercus taurica	PO	[146, 133]	*Nasitrema sunameri**	PO	[309, 314, 306, 146, 133]
						*Pharurus asiaeorientalis**	PO, YR	[146, 133]	*Odhneriella elongata*	PO	[210, 314]
						*Pharurus sunameri**	PO	[310, 306, 146, 133]	*Synthesium elongatum*	PO	[210, 76, 306, 133]
						Stenurus auditivus	PO	[94, 76, 306]	*Synthesium nipponicum*	PO	[146, 133]
						*Stenurus nanjingensis**	PO	[133]	*Synthesium tursionis*	IO	[121, 133]
Phocoena sinus Norris and McFarland, 1958. Vaquita											
-	-	-	-	-	-	-	-		*Synthesium tursionis*	PO	[150]

Acanthocephala	Locality	Ref.	Cestoda	Locality	Ref.	Nematoda	Locality	Ref.	Trematoda	Locality	Ref.
				Phocoena phocoena (Linnaeus, 1758). Harbor porpoise							
Bolbosoma capitatum	AO	[116]	*Diphyllobothrium lanceolatum*	AO, BlS	[76, 77, 306, 229]	*Anisakis simplex* s.l.*	AO, PO, BlS, BaS	[31, 76, 77, 73, 306, 115, 53, 236, 170, 24, 41, 45, 125, 116, 1, 155, 153]	*Brachycladium nipponicum*	PO	[306]
*Corynosoma alaskensis**	PO	[117, 171, 306]	*Diphyllobothrium latum**	BlS	[76, 77, 306, 38]	*Anisakis simplex* s.s.	AO, PO	[182]	*Campula oblonga*	AO, PO, Arc, MS, BlS, BaS	[31, 219, 76, 77, 51, 314, 171, 306, 115, 229, 237, 53, 236, 170, 24, 41, 45, 249, 104, 116, 155, 109, 153]
*Corynosoma semerme**	BlS	[76, 77, 306]	*Diphyllobothrium phocarum**	PO	[240, 171]	*Anisakis typica*	AO, BlS	[31, 76, 77, 73, 306, 184]	*Opisthorchis tenuicollis**	AO, Arc	[76]
*Corynosoma strumosum**	BlS	[76, 77, 306]	*Diphyllobothrium stemmacephalum*	AO, BlS	[31, 76, 77, 312, 79, 306, 53, 229, 41, 45, 125, 116, 38]	*Phocascaris* sp.*	AO	[41]	*Pholeter gastrophilus*	AO, BlS, BaS	[31, 76, 77, 306, 92, 115, 229, 24, 125, 116, 38, 153]
			*Monorygma grimaldii**	AO	[17, 155]	*Pseudoterranova decipiens**	AO	[76, 32, 306]	*Synthesium mironovi*	PO	[69]
			*Phyllobothrium delphini**	AO	[17, 155]	*Porrocaecum* sp.*	AO	[256]	*Synthesium nipponicum*	PO	[51, 171, 69, 170]

Table 2. (Continued)

Acanthocephala	Locality	Ref.	Cestoda	Locality	Ref.	Nematoda	Locality	Ref.	Trematoda	Locality	Ref.
			Tetrabothrius sp.	AO	[41]	*Halocercus invaginatus*	AO, PO, Arc, BlS BaS	[94, 76, 77, 171, 306, 13, 115, 24, 195, 249, 1, 116, 38, 154, 155, 153, 214]			
						Halocercus kirbyi	PO	[190]			
						Halocercus taurica	AO, PO, BlS	[76, 77, 306, 13, 116, 38, 214]			
						Pseudalius inflexus	AO, PO, BlS, BaS	[31, 94, 76, 77, 304, 306, 13, 115, 53, 236, 24, 45, 249, 116, 154, 153]			
						Stenurus minor	AO, PO, BlS, BaS	[31, 76, 77, 171, 306, 13, 115, 237, 53, 236, 24, 45, 249, 116, 38, 154, 155, 153, 214]			
						Torynurus convolutus	AO, PO, BlS BaS	[31, 76, 77, 171, 306, 13, 115, 236, 24, 45, 249, 116, 154, 153]			
						*Hysterothylacium aduncum**	AO	[125]			
						Crassicauda sp.	AO, PO	[69, 170, 116]			
Phocoena spinipinnis Burmeister, 1865. Burmeister's porpoise											
*Corynosoma australe**	AO, PO	[287, 33]	*Monorygma grimaldii**	§	[17]	*Anisakis simplex* s.l.*	AO, PO	[287, 21, 33]	*Braunina cordiformis*	PO	[244]
Corynosoma cetaceum	AO, PO	[287, 57, 33]	*Phyllobothrium delphini**	§	[17]	*Anisakis typica*	PO	[244]	*Nasitrema globicephalae*	PO	[244]
						Pseudalius inflexus	AO, PO	[57, 244, 280, 50]	*Pholeter gastrophilus*	AO, PO	[57, 244, 33]
						Pseudoterranova sp.	PO	[287]	*Synthesium tursionis*	PO	[287, 244, 280]
						*Stenurus australis**	AO, PO	[280, 190, 320, 201]			
						Stenurus minor	AO, PO	[57, 280]			
Phocoenoides dalli (True, 1885). Dall's porpoise											
-	-	-	*Monorygma grimaldii**	§	[17]	*Anisakis simplex* s.l.*	PO	[138, 61, 171, 115, 132, 170]	*Campula oblonga*	PO	[309, 51, 314, 61, 193, 306, 70, 56, 132, 170]

Acanthocephala	Locality	Ref.	Cestoda	Locality	Ref.	Nematoda	Locality	Ref.	Trematoda	Locality	Ref.
			*Phyllobothrium delphini**	PO	[17, 300]	*Crassicauda boopis*	PO	[279]	*Nasitrema dalli*	PO	[309, 314, 61, 306, 71, 132]
			Tetrabothrius forsteri	PO	[311, 306]	*Halocercus dalli**	PO	[310, 306, 56, 132]	*Nasitrema delphini*	PO	[62]
						Halocercus kirbyi	PO	[76, 171, 306, 115]	*Nasitrema globicephalae*	PO	[62]
						Halocercus pingi	PO	[190]	*Synthesium nipponicum*	PO	[309, 314, 306]
						Placentonema sp.	§	[132]			
						Stenurus minor	PO	[136, 193, 171, 62, 71, 132]			
						*Stenurus truei**	PO	[190]			
						*Stenurus yamagutii**	PO	[190]			
						*Torynurus dalli**	PO	[306, 13, 71, 132, 170]			
Family Physeteridae *Physeter macrocephalus* Linnaeus, 1758. Sperm whale											
Bolbosoma brevicolle	AO	[31, 76, 306, 84]	*Diphyllobothrium* sp.	Ant	[70]	*Anisakis pegreffii*	§	[182]	*Brachycladium curilense*	PO	[76, 314, 306, 84]
Bolbosoma capitatum	AO, MS, Ant	[31, 70, 76, 306, 84]	*Diplogonoporus* sp.	§	[84]	*Anisakis physeteris*	AO, PO, Ant, MS, SA	[31, 76, 172, 138, 73, 171, 306, 115, 170, 70, 200, 147, 181, 201, 33, 48]			
Bolbosoma tuberculata	AO, IO	[266, 273]	*Hexagonoporus calyptocephalus**	PO	[306, 85]	*Anisakis simplex* s.l.*	AO, PO, SA	[76, 172, 138, 73, 171, 306, 115, 170]			
Bolbosoma turbinella	PO	[84, 306]	*Hexagonoporus giganticus**	Ant	[264, 306, 273]	*Crassicauda anthonyi*	AO	[229]			

Table 2. (Continued)

Acanthocephala	Locality	Ref.	Cestoda	Locality	Ref.	Nematoda	Locality	Ref.	Trematoda	Locality	Ref.
Corynosoma bullosum[*]	AO, Ant	[70, 33]	*Hexagonoporus physeteris*[*]	PO	[76, 306, 84]	*Placentonema gigantissimum*[*]	AO, PO	[76, 245, 263, 84, 171, 306, 115, 70, 170]			
Corynosoma curilense[*]	PO	[76, 306, 84]	*Monorygma grimaldii*[*]	PO, Ant	[269, 306, 115, 70]	*Stenurus ovatus*	AO	[229]			
Corynosoma strumosum[*]	PO	[84, 307]	*Multiductus physeteris*[*]	PO	[115]						
			Phyllobothrium delphini[*]	AO, PO, Ant, MS, SA	[31, 76, 172, 245, 263, 304, 84, 270, 307, 171, 115, 229, 70, 116, 191, 47]						
			Priapocephalus grandis	AO, Ant	[243, 177, 52, 84, 306, 115, 229, 70]						
			Scolex pleuronectis[*]	PO	[177, 306, 9]						
			Tentacularia sp.[*]	§	[273]						
			Tetrabothrius affinis	AO, PO, Ant	[177, 84, 306, 115]						
			Tetrabothrius curilensis[*]	Ant	[76, 306, 277, 70]						
			Tetrabothrius wilsoni	Ant	[243, 54, 70]						
			Trigonocotyle sp.	PO	[76]						
Family Platanisitdae											
Platanista gangetica (Lebeck, 1801). South Asian river dolphin											
-	-	-	-	-	-	*Contracaecum lobulatum*[*]	GR	[31, 76, 216 65]	*Campula oblonga*	GR	[31, 76, 65]

Acanthocephala	Locality	Ref.	Cestoda	Locality	Ref.	Nematoda	Locality	Ref.	Trematoda	Locality	Ref.
Family Pontoporiidae											
Pontoporia blainvillei (Gervais and d'Orbigny, 1844). Franciscana											
Bolbosoma turbinella	AO	[10]	*Monorygma grimaldii**	AO	[17]	*Anisakis simplex* s.l.*	AO	[18, 21]	*Pholeter gastrophilus*	AO	[18]
*Corynosoma australe**	AO	[10, 20]	*Phyllobothrium delphini**	AO	[17]	*Anisakis typica*	AO	[139, 22, 10, 192]	*Synthesium pontoporiae**	AO	[18, 228, 10, 176, 35, 174, 173]
Corynosoma cetaceum	AO	[254, 139, 18, 10, 19, 20]				*Procamallanus* sp.*	AO	[22]			
Family Ziphiidae											
Berardius bairdii Stejneger, 1883. Baird's beaked whale											
-	-	-	*Phyllobothrium delphini**	PO	[245, 171]	*Anisakis simplex* s.l.*	PO	[172, 171, 306]	*Oschmarinella sobolevi*	PO	[306]
						Crassicauda giliakiana	PO	[306, 279]			
						*Halocercus hyperoodoni**	PO	[190]			
Hyperoodon ampullatus (Forster, 1770). Northern bottlenose whale											
Bolbosoma balaenae	AO	[76, 306]	*Strobilocephalus triangularis*	AO	[23, 31, 76, 306, 229]	*Anisakis physeteris*	§	[73, 84]	*Oschmarinella sobolevi*	AO	[260, 76, 314]
Bolbosoma turbinella	AO	[31, 76, 78, 270, 306]				*Anisakis simplex* s.l.*	AO	[31, 76, 73]			
						Anisakis typica	§	[137]			
						Crassicauda giliakiana	§	[76]			
Hyperoodon planifrons Flower, 1882. Southern bottlenose whale											
-	-	-	-	-	-	*Crassicauda giliakiana*	Ant	[306]	-	-	-
Mesoplodon bidens (Sowerby, 1804). Sowerby's beaked whale											
*Bolbosoma vasculosum**	AO, MS$^{\infty}$	[295, 76, 306]	*Phyllobothrium delphini**	AO, MS$^{\infty}$	[76, 263, 303, 306, 229]	*Anisakis simplex* s.l.*	AO	[31, 76, 73, 306, 115]	-	-	-

Table 2. (Continued)

Acanthocephala	Locality	Ref.	Cestoda	Locality	Ref.	Nematoda	Locality	Ref.	Trematoda	Locality	Ref.
			Strobilocephalus triangularis	AO	[76, 229]						
			Tetrabothrius forsteri	MS$^{\infty}$	[23, 31, 76, 306, 229]						
$^{\infty}$Single record of the host species in the Mediterranean Sea											
Mesoplodon bowdoini Andrews, 1908. Andrews' beaked whale											
-	-	-	-	-	-	*Anisakis nascettii*	PO	[185]	-	-	-
						Anisakis ziphidarum	PO	[185]			
Mesoplodon densirostris (Blainville, 1817). Blainville's beaked whale											
*Bolbosoma vasculosum**	AO	[318, 55]	*Phyllobothrium delphini**	AO	[318]	*Anisakis simplex* s.l.*	AO	[318, 116]			
			Tetrabothrius sp.	AO	[116]	*Anisakis ziphidarum*	SA	[181]			
						Crassicauda sp.	PO	[279]			
Mesoplodon europaeus (Gervais, 1855). Gervais' beaked whale											
-	-	-	*Monorygma grimaldii**	§	[17]	*Anisakis simplex* s.s.	AO	[48]	*Brachycladium parvulum**	AO	[108]
			*Phyllobothrium delphini**	§	[17]	*Anisakis ziphidarum*	AO	[181, 48]	*Cetitrema meadi**	AO	[64]
Mesoplodon grayi von Haast, 876. Gray's beaked whale											
-	-	-	*Phyllobothrium delphini**	AO	[33]	*Anisakis nascettii*	PO	[185]	-	-	-
			*Scolex pleuronectis**	AO	[33]	*Anisakis ziphidarum*	§	[181]			
Mesoplodon hectori (Gray, 1871). Hector's beaked whale											
Bolbosoma sp.	AO	[46]	*Tetrabothrius hobergi**	AO	[205]	*Anisakis* sp.	AO	[46]	*Braunina cordiformis*	AO	[46]
Mesoplodon layardii (Gray, 1865). Layard's beaked whale											
-	-	-	-	-	-	*Anisakis nascettii*	PO	[185]	-	-	-
						Anisakis berlandi	SA	[182]			
						Anisakis ziphidarum	SA	[185]			

Acanthocephala	Locality	Ref.	Cestoda	Locality	Ref.	Nematoda	Locality	Ref.	Trematoda	Locality	Ref.
						Mesoplodon mirus True, 1913. True's beaked whale					
-	-	-	*Monorygma grimaldii**	§	[17]	*Anisakis nascettii*	SA	[185]	-	-	-
			*Phyllobothrium delphini**	§	[17]	*Anisakis simplex* s.l.*	SA	[115]			
						Crassicauda anthonyi	AO	[49, 306]			
						Mesoplodon stejnegeri True, 1885. Stejneger's beaked whale					
-	-	-	*Tetrabothrius forsteri*	Arc	[259]	*Crassicauda* sp.	PO	[279]	*Oschmarinella macrorchis*	PO	[87]
						Ziphius cavirostris G. Cuvier, 1823. Cuvier's beaked whale					
*Bolbosoma vasculosum**	MS	[100]	*Monorygma grimaldii**	AO	[55]	*Anisakis pegreffii*	MS	[182, 39]	*Oschmarinella albamarina*	PO	[207]
			*Phyllobothrium delphini**	AO	[194, 55]	*Anisakis physeteris*	MS	[162, 39]			
			*Scolex pleuronectis**	MS	[9]	*Anisakis simplex* s.l.*	PO	[147]			
			Tetrabothrius sp.	MS	[100]	*Anisakis typica*	MS	[229]			
						Anisakis ziphidarum	AO, MS, SA	[181, 55]			
						Crassicauda anthonyi	AO, PO, MS	[49, 90, 306, 194, 55, 207]			
						Crassicauda boopis	MS	[31, 162]			
						Crassicauda crassicauda	AO, MS	[76, 162, 115]			

Table 2. (Continued)

No helminths reported from the following cetacean species:

Family Balaenidae, North Pacific right whale, *Eubalaena japonica* (Lacépède, 1818).

Family Balaenopteridae, Antarctic minke whale, *Balaenoptera bonaerensis* Burmeister, 1867.

Omura's whale, *Balaenoptera omurai* Wada, Oishi and Yamada, 2003.

Family Delphinidae, Australian snubfin dolphin, *Orcaella heinsohni* Beasley, Robertson and Arnold, 2005.

Atlantic humpback dolphin, *Sousa teuszii* (Kükenthal, 1892).

Indian Ocean humpback dolphin, *Sousa plumbea* (G. Cuvier, 1829).

Australian humpback dolphin, *Sousa sahulensis.*

Family Lipotidae, Baiji, *Lipotes vexillifer* (Miller, 1918).

Family Phocoenidae, Narrow-ridged finless porpoise, *Neophocaena asiaeorientalis* (Pilleri and Gihr, 1972).

Family Ziphiidae, Arnoux' beaked whale, *Berardius arnuxii* Duvernoy, 1851.

Longman's beaked whale, *Indopacetus pacificus* (Longman, 1926).

Hubbs' beaked whale, *Mesoplodon carlhubbsi* Moore, 1963.

Ginkgo-toothed beaked whale, *Mesoplodon ginkgodens* Nishiwaki and Kamiya, 1958.

Deraniyagala's beaked whale, *Mesoplodon hotaula* Deraniyagala, 1963.

Perrin's beaked whale, *Mesoplodon perrini* Dalebout, Mead, Baker, Baker and van Helden, 2002.

Pygmy beaked whale, *Mesoplodon peruvianus* Reyes, Mead and Van Waerebeek, 1991.

Spade-toothed whale, *Mesoplodon traversii* (Gray, 1874).

Shepherd's beaked whale, *Tasmacetus shepherdi* Oliver, 1937.

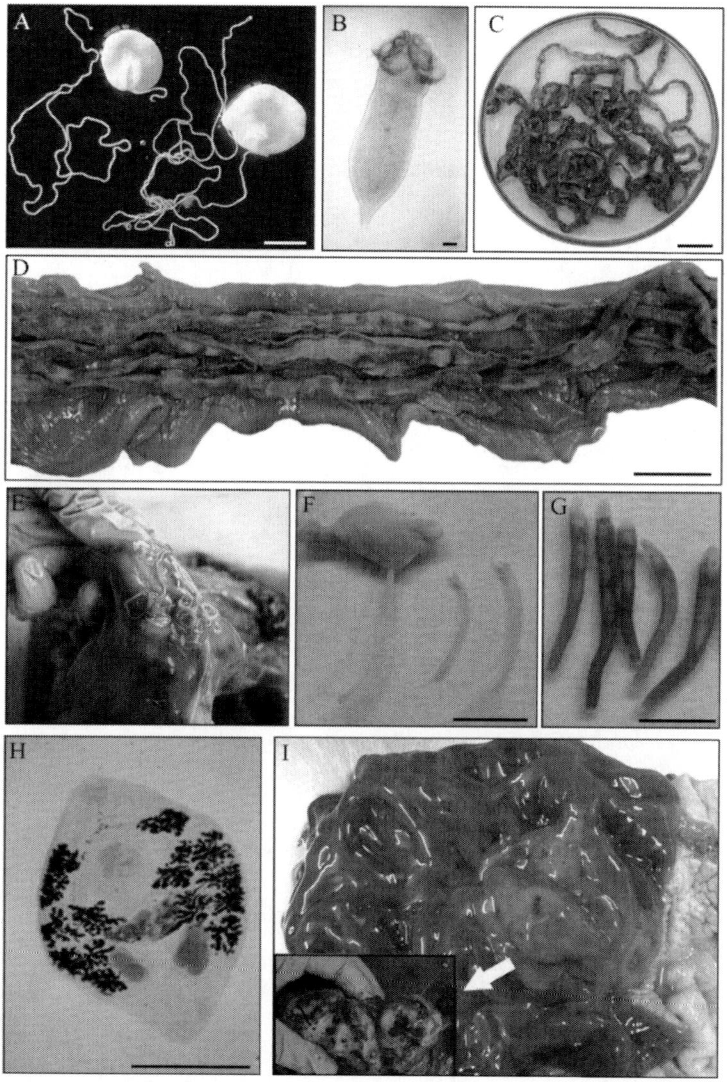

Figure 1. Representatives of helminth species in cetaceans. A) *Monorygma grimaldii* (Cestoda) and its capsule. Scale bar: 1 cm. B) *Scolex pleuronectis* (Cestoda). Scale bar: 0.1 mm. C) and D) *Diphyllobothrium* sp. (Cestoda) in the intestine of a bottlenose dolphin, *Tursiops truncatus*. Scale bars: 2 cm. E) *Anisakis simplex* sensu lato (Nematoda) in the stomach of a striped dolphin, *Stenella coeruleoalba*. F) Two specimens of *Bolbosoma capitatum* (Acanthocephala) and one embedded in host tissue. Scale bar: 1 cm. G) *Oschmarinella rochebruni* (Digenea). Scale bar: 1 cm. H) *Pholeter gastrophilus* (Digenea). Scale bar: 1 cm. I) Cyst of *P. gastrophilus* in the stomach of a striped dolphin. Pictures from the University of Valencia.

ACKNOWLEDGMENTS

We are grateful to Dr. Rodney Bray from the Natural History Museum (London, UK), who kindly clarified some taxonomic doubts on the validity of certain cestode species. This study was supported by projects CGL/2009/07465 and CGL2012-39545 from the Ministry of Economy and Competitiveness, Spain, and PROMETEOII/2015/018 of the Valencian Government, Spain.

REFERENCES

Abollo, E., López, A., Gestal, C., Benavente, P., and Pascual, S. (1998). Macroparasites in cetaceans stranded on the northwestern Spanish Atlantic coast. *Diseases of Aquatic Organisms,* 32, 227 – 231 [1].

Abollo, E., and Pascual, S. (2002). SEM study of *Anisakis brevispiculata* Dollfus, 1966 and *Pseudoterranova ceticola* (Deardoff and Overstreet, 1981) (Nematoda: Anisakidae), parasites of the pygmy sperm whale *Kogia breviceps*. *Scientia Marina,* 66, 249 – 255 [2].

Abril, E., Almor, P., Raga, J. A., and Duguy, R. (1987). Parasitisme par *Anisakis typica* (Diesing, 1860) chez le dauphin bleu et blanc (*Stenella coeruleoalba*) dans le nord-est Atlantique. *Bulletin de la Société Zoologique de France,* 111, 131 – 133 [3].

Abril, E., Balbuena, J. A., and Raga, J. A. (1991). A new species of the genus *Zalophotrema* (Digenea: Campulidae), *Zalophotrema atlanticum* n. sp. from the liver of the striped dolphin *Stenella coeruleoalba* (Meyen, 1833) (Cetacea: Delphinidae) in Atlantic waters. *Systematic Parasitology,* 18, 133 – 138 [4].

Adams, A. M., Hoberg, E. P., McAlpine, D. F., and Clayden, S. L. (1998). Occurrence and morphological comparison of *Campula oblonga* (Digenea: Campulidae), including a report from an atypical host, the thresher shark, *Alopias vulpinus*. *Journal of Parasitology,* 84, 435 – 438 [5].

Aguilar-Aguilar, R., Delgado-Estrella, A., and Moreno-Navarrete, A. G. (2010). New host record for nematodes in a stranded short-snouted spinner dolphin *Stenella clymene* (Cetacea: Delphinidae) from the Mexican Caribbean coast. *Helminthologia,* 47, 136 – 138 [6].

Aguilar-Aguilar, R., Moreno-Navarrete, Salgado-Maldonado, G., and Villa-Ramírez, B. (2001). Gastrointestinal helminths of spinner dolphins *Stenella longirostris* (Gray, 1828) (Cetacea: Delphinidae) stranded in La Paz Bay, Baja California Sur, Mexico. *Comparative Parasitology,* 68, 272 – 274 [7].

Agustí, C., Aznar, F. J., and Raga, J. A. (1999). Intestinal helminth communities of Mediterranean striped dolphin. Paper presented at the *13th Biennial Conference on the Biology of Marine Mammals*. Hawai, November 28 - December 3 [8].

Agustí, C., Aznar, F. J., and Raga, J. A. (2005). Tetraphyllidean plerocercoids from Western Mediterranean cetaceans and other marine mammals around the world: A comprehensive morphological analysis. *Journal of Parasitology,* 91, 83 – 92 [9].

Andrade, A., Pinedo, C., and Pereira, J. (1997). The gastrointestinal helminths of the Franciscana, *Pontoporia blainvillei,* in Southern Brazil. *Reports of the International Whaling Commission,* 47, 669 – 673 [10].

Andrade, A., Pinedo, C., and Barreto, A. S. (2001). Gastrointestinal parasites and prey items from a mass stranding of false killer whales, *Pseudorca crassidens,* in Rio Grande do Sul, Southern Brazil. *Revista Brasileira de Biologia,* 61, 55 – 61 [11].

Appy, R. G. (1974). *The air sinus flukes, genus Nasitrema (Trematoda: Nasitrematidae), in porpoises from the Eastern Pacific*. Master thesis, California State University [12].

Arbelo, M., Espinosa de los Monteros, A., Herraéz, P., Andrada, M., Sierra, E., Rodríguez, F., Jepson, P. D., and Fernández, A. (2013). Pathology and causes of death of stranded cetaceans in the Canary Islands (1999 – 2005). *Diseases of Aquatic Organisms,* 193, 87 – 99.

Arnold, P. W., and Gaskin, D. E. (1975). Lungworms (Metastrongyloidea: Pseudaliidae) of harbor porpoise *Phocoena phocoena* (L. 1758). *Canadian Journal of Zoology,* 53, 713 – 735 [13].

Arvy, L. (1973/74). The kidney, renal parasites and renal secretion in Cetaceans. *Investigations on Cetacea,* 5, 231 – 310 [14].

Arvy, L. (1982). Phoresis and parasitism in cetaceans, a review. *Investigations on Cetacea,* 14, 233 – 335 [15].

Aytemiz, I., Dede, A., Danyer, E., and Tonay, A. (2012). Morphological identification of parasites found in the stomach contents of bycaught striped dolphins (*Stenella coeruleoalba*) from Turkish Eastern Mediterranean Sea coast. *Journal of the Black Sea/Mediterranean Environment,* 18, 238 – 245 [16].

Aznar, F. J., Agustí, C., Littlewood, D. T. J., Raga, J. A., and Olson, P. D. (2007). Insight into the role of cetaceans in the life cycle of the tetraphyllideans (Platyhelminthes: Cestoda). *International Journal for Parasitology,* 37, 243 – 255 [17].

Aznar, F. J., Balbuena, J. A., Fernández, M., and Raga, J. A. (2001a). Living together: the parasites of marine mammals. In P. Evans, and J. A. Raga (Eds.), *Marine mammals: Biology and Conservation* (pp. 385 – 421). New York: Kluwer Academic/Plenum Publishers.

Aznar, F. J., Balbuena, J. A., and Raga, J. A. (1994). Helminth communities of *Pontoporia blainvillei* (Cetacea: Pontoporiidae) in Argentinian waters. *Canadian Journal of Zoology,* 72, 702 – 706 [18].

Aznar, F. J., Bush, A. O., Balbuena, J. A., and Raga, J. A. (2001b). *Corynosoma cetaceum* in the stomach of Franciscanas, *Pontoporia blainvillei* (Cetacea): An exceptional case of habitat selection by an acanthocephalan. *Journal of Parasitology,* 87, 536 – 541 [19].

Aznar, F. J., Fernández, M., and Balbuena, J. A. (2010). Why we should care about the parasite fauna of cetaceans: A plea for integrative studies. In C. A. Murray (Ed.), *Whales and Dolphins: Behavior, Biology and Distribution* (pp. 115 – 127). New York: Nova Science Publishers, Inc.

Aznar, F. J., Hernández-Orts, J., Suárez, A. A., Gacía-Varela, M., Raga, J. A., and Cappozzo, H. L. (2012). Assessing host–parasite specificity through coprological analysis: a case study with species of *Corynosoma* (Acanthocephala: Polymorphidae) from marine mammals. *Journal of Helminthology,* 86, 156 – 164 [20].

Aznar, F. J., Herreras, M. V., Balbuena, J. A., and Raga, J. A. (2003). Population structure and habitat selection by *Anisakis simplex* in 4 odontocete species from Northern Argentina. *Comparative Parasitology,* 70, 66 – 71 [21].

Aznar, F. J., Perdiguero, D., Pérez del Olmo, A., Repullés, A., Agustí, C., and Raga, J. A. (2005). Changes in epizoic crustacean infestations during cetacean die-offs: the mass mortality of Mediterranean striped dolphins *Stenella coeruleoalba* revisited. *Diseases of Aquatic Organisms,* 67, 239 – 247.

Aznar, F. J., Raga, J. A., Corcuera, J., and Monzón, F. (1995). Helminths as biological tags for Franciscana (*Pontoporia blaninvillei*) (Cetacea, Pontoporiidae) in Argentinian and Uruguayan waters. *Mammalia,* 59, 427 – 435 [22].

Baer, J. G. (1932). Contribution à l'étude des Cestodes de Cétacés. *Revue Suisse de Zoologie,* 39, 393 – 418 [23].

Baker, J. R., and Martin, A. R. (1992). Causes of mortality and parasites and incidental lesions in harbour porpoises (*Phocoena phocoena*) from British waters. *The Veterinary Record,* 130, 554 – 558 [24].

Balbuena, J. A., Aznar, F. J., Fernández, M., and Raga, J. A. (1995). Parasites as indicators of social structure and stock identity of marine mammals. *Whales, Seals, Fish and Men,* 4, 133 – 139 [25].

Balbuena, J. A., and Raga, J. A. (1993a). Diphyllobothriids (Cestoda: Pseudophyllidea) from the long-finned pilot whale, *Globicephala melas* (Traill, 1809) off the Faroe Island with comments on the taxonomy of *Diphyllobothrium* Cobbold, 1858 species from cetaceans. *Systematic Parasitology,* 26, 117 – 126 [26].

Balbuena, J. A., and Raga, J. A. (1993b). Intestinal helminth communities of the long-finned pilot whale (*Globicephala melas*) off the Faroe Islands. *Parasitology,* 106, 327 – 333 [27].

Balbuena, J. A., and Raga, J. A. (1994). Intestinal helminths as indicators of segregation and social structure of pods of long-finned pilot whales (*Globicephala melas*) off the Faroe Islands. *Canadian Journal of Zoology,* 72, 443 – 448 [28].

Balbuena, J. A., Raga, J. A., and Abril, E. (1989). Redescription of *Odhneriella subtila* (Skrjabin, 1959) (Digenea: Campulidae) from the intestine of *Globicephala melaena* (Traill, 1809) (Cetacea: Delphinidae) off the Faroe Islands (North-East Atlantic). *Systematic Parasitology,* 14, 31 – 36 [29].

Baylis, H. A. (1929). Parasitic Nematoda and Acanthocephala collected in 1925-1927. *Discovery Reports,* 1, 541 – 559 [30].

Baylis, H. A. (1932). A list of worms parasitic in Cetacea. *Discovery Reports,* 6, 393 – 418 [31].

Berland, B. (1960). Om rundmarklarver i fisk. *Saertrykk av Naturen,* 8, 455 – 465 [32].

Berón-Vera, B., Crespo, E. A., and Raga, J. A. (2008). Parasites in stranded cetaceans of Patagonia. *Journal of Parasitology,* 94, 946 – 948 [33].

Berón-Vera, B., Crespo, E. A., Raga, J. A., and Fernández, M. (2007). Parasite communities of common dolphins (*Delphinus delphis*) from Patagonia: the relation with host distribution and diet and comparison with sympatric hosts. *Journal of Parasitology,* 93, 1056 – 1060 [34].

Berón-Vera, B., Fernández, M., Crespo, E. A., Aznar, F. J., and Raga, J. A. (2003). *Hadwenius* spp. (Digenea: Campulidae) off Patagonia, Argentina. *Bulletin of the Scandinavian Society for Parasitology,* 30, 12 – 13 [35].

Berón-Vera, B., Pedraza, S. N., Raga, J. A., Gil de Pertierra, A., Crespo, E. A., Alonso, M. K., and Goodall, R. N. P. (2001). Gastrointestinal helminths of Commerson's dolphins *Cephalorhynchus commersonii* from central Patagonia and Tierra del Fuego. *Diseases of Aquatic Organisms,* 47, 201 – 208 [36].

Beverley-Burton, B. (1978). Helminths of the alimentary track from a stranded herd of the Atlantic white-sided dolphin, *Lagenorhynchus acutus*. *Journal of Fisheries Research Board of Canada,* 35, 1356 – 1359 [37].

Birkun, A. (2002). Natural mortality factors affecting cetaceans in the Black Sea. In G. Notarbartolo di Sciara (Ed.), *Cetaceans of the Mediterranean and Black Seas: state of knowledge and conservation strategies* (pp. 1 – 13). Monaco: ACCOBAMS Secretariat [38].

Blazekovic, K., Lepen Pleic, I., Đuras, M., Gomercic, T., and Mladineo, I. (2015). Three *Anisakis* spp. isolated from toothed whales stranded along the eastern Adriatic Sea coast. *International Journal for Parasitology,* 45, 17 – 31 [39].

Bowie, J. Y. (1984). Parasites from an Atlantic bottle-nosed dolphin (*Tursiops truncatus*), and a revised checklist of parasites of this host. *New Zealand Journal of Zoology,* 11, 395 – 398 [40].

Brattey, J., and Stenson, G. B. (1993). Host specificity and abundance of parasitic nematodes (Ascaridoidea) from the stomachs of five phocids species from Newfoundland and Labrador. *Canadian Journal of Zoology,* 71, 2156 – 2166.

Brattey, J., and Stenson, G. B. (1995). Helminth parasites of the alimentary tract of the harbour porpoise, *Phocoena phocoena* (L.) from Newfoundland and Labrador. *Journal of the Helminthological Society of Washington,* 62, 209 – 216 [41].

Braun, M. (1902). Ueber *Distoma goliath* P.J. V.Ben. 1858. *Centralblatt Bacteriologie und Parasitenkunde. ABTH I,* 32, 800 – 803 [42].

Brieva, L. M., and Oporto, J. A. (1991). Prevalencia e intensidad de la infección por trematodos del género *Nasitrema* en el delfín chileno *Cephalorhynchus eutropia* (Cetacea: Delphinidae). *Archivos de Medicina Veterinaria, Chile,* 23, 97 – 100 [43].

Briscoe, A. G., Bray, R. A., Brabec, J. and Littlewood, D. T. J. (2016) The mitochondrial genome and ribosomal operon of *Brachycladium goliath* (Digenea: Brachycladiidae) recovered from a stranded minke whale. *Parasitology International,* 65, 271 – 275 [44].

Brosens, L., Jauniaux, T., Siebert, U., Benke, H., and Coignoul, F. (1996). Observations on the helminths of harbour porpoises (*Phocoena phocoena*) and common guillemots (*Uria aalgae*) from the Belgian and German coasts. *Veterinary Record,* 139, 254 – 257 [45].

Capozzo, H. L., Negri, M. F., Mahler, B., Lía, V. V., Martínez, P., Gianggiobe, A., and Saubidet, A. (2005). Biological data on two Hector's beaked whales, *Mesoplodon hectori*, stranded in Buenos Aires Province, Argentina. *Latin American Journal of Aquatic Mammals,* 4, 113 – 128 [46].

Carvalho, V. L., Leal Bevilaqua, C. M., Mayo Iñiguez, A., Mathews-Cascon, H., Bezerra Ribeiro, F., Bezerra Pessoa, L. M., Oliveira de Meirelles, A. C., Gomes Borges, J. C., Marigo, J., Soares, L., and de Lima Silva, F. J. (2010). Metazoan parasites of cetaceans off the northeastern coast of Brazil. *Veterinary Parasitology,* 173, 116 – 122 [47].

Cavallero, S., Nadler, S. A., Paggi, L., Barros, N. B., and D'Amelio, S. (2011). Molecular characterization and phylogeny of anisakid nematodes from cetaceans from southeastern Atlantic coasts of USA, Gulf of Mexico, and Caribbean Sea. *Parasitology Research,* 108, 781 – 792 [48].

Chabaud, A. G. (1962). Description de *Crassicauda anthonyi* n. sp., nematode parasite renal de *Mesoplodon mirus* True. *Bulletin du Muséum National d'Histoire Naturelle, Paris,* 34, 397 – 403 [49].

Chavera, A., Cabrera, R., and Tantalean, M. (2010). Descripción de lesiones pulmonares por *Pseudalius inflexus* en la marsopa *Phocaena spinipinnis*. *Revista Peruana de Biología,* 17, 261 – 264 [50].

Ching, H. L., and Robinson, E. S. (1959). Two campulid trematodes from a new host, the harbor porpoise. *Journal of Parasitology,* 45, 181 [51].

Clark, R. (1956). Sperm whales of the Azores. *Discovery Reports,* 28, 239 – 298 [52].

Clausen, B., and Andersen, S. (1988). Health status of harbour porpoise (*Phocoena phocoena*) from Danish waters. *Danish Review of Game Biology,* 13, 1 – 20 [53].

Cockrill, W. R. (1960). Pathology of Cetacea, a veterinary study on whales. Part I; II. *British Veterinary Journal,* 116, 133 – 144 [54].

Colón-Llavina, M. M., Mignucci-Giannoni, A., Mattiucci, S., Paoletti, M., Nascetti, G., and Williams, E. H. (2009). Additional records of metazoan parasites from Caribbean marine mammals, including genetically identified anisakid nematodes. *Parasitology Research,* 105, 1239 – 1252 [55].

Committee on Taxonomy. (2014). List of marine mammal species and subspecies. Society for Marine Mammalogy. Accessed through: www.marinemammalscience.org on 2015 – 08 – 01.

Conlongue, G. J., Ogden, J. A., and Foreyt, W. J. (1985). Parasites of the Dall's porpoise (*Phocoenoides dalli*). *Journal of Wildlife Diseases,* 21, 160 – 166 [56].

Corcuera, J., Monzón, F., Aguilar, A., Borrell, A., and Raga, J. A. (1995). Life history data, organochlorine pollutants and parasites from eight Burmeister's porpoises, *Phocoena spinipinnis*, caught in Northern Argentine waters. *Reports of the International Whaling Commission,* 16, 365 – 372 [57].

Cornaglia, E. Rebora, L. Gili, C., and Di Guardo, G. (2000). Histopathological and immunohistochemical studies on cetaceans found stranded on the coast of Italy between 1990 and 1997. *Journal of Veterinary Medicine Series A,* 47, 129 – 142 [58].

Cowan, D. F. (1967). Helminth parasites of the pilot whale *Globicephala melaena* (Traill 1809). *Journal of Parasitology,* 53, 166 – 167 [59].

Dailey, M. D. (1971a). A new species of *Hunterotrema* (Digenea: Campulidae) from the Amazon river dolphin (*Inia geoffrensis*). *Bulletin of the Southern California Academy of Sciences,* 70, 79 – 80 [60].

Dailey, M. D. (1971b). Distribution of helminths in the Dall porpoise (*Phocoenoides dalli* True). *The Journal of Parasitology,* 57, 1348 [61].

Dailey, M. D. (1978). Baseline data on parasites from marine mammals in Southern California. In *Marine mammals and seabird survey of the Southern California Bight area. Vol. III.* Principal investigator's reports. Book 1. Pinnipedia, Cetacea and Parasitology (pp. 442 – 472). Santa Cruz: University of California [62].

Dailey, M. D. (1985). Diseases of Mammalia: Cetacea. In O. Kinne (Ed.), *Diseases of Marine Animals* (pp. 805 – 847). Hamburg: Biologische Anstalt Helgoland [63].

Dailey, M. D. (2007). A new species of Digenea (Trematoda: Brachycladiidae) from the Gervais' Beaked whale, *Mesoplodon europaeus*, with comments on other cetacean liver flukes. *Comparative Parasitology,* 74, 229 – 232 [64].

Dailey, M. D., and Brownell, R. L. Jr. (1972). A checklist of marine mammal parasites. In S. R. Ridgway (Ed.), *Mammals of the sea. Biology and Medicine* (pp. 528 – 589). Springfield: Charles C. Thomas [65].

Dailey, M. D., Demaree, R. S., and Critchfield, R. L. (2002). *Galactosomum stelleri* sp. n. (Trematoda: Heterophyidae) from the Northern Sea-Lion, *Eumetopias jubatus* (Schreber, 1776) (Carnivora: Otariidae). *Comparative Parasitology,* 69, 58 – 61.

Dailey, M. D., Gulland, F. M. D., Lowenstine, L. J., Silvagni, P., and Howard, D. (2000). Prey, parasites and pathology associated with the mortality of a juvenile gray whale (*Eschrichtius robustus*) stranded along the northern California coast. *Diseases of Aquatic Organisms,* 42, 111 – 117 [66].

Dailey, M. D., and Otto, K. A. (1982). *Parasites as biological indicators of the distributions and diets of marine mammals common to the Eastern Pacific*. California: National Oceanic and Atmospheric Administration [67].

Dailey, M. D., and Perrin, W. F. (1973). Helminth parasites of porpoises of the genus *Stenella* in the Eastern Tropical Pacific, with descriptions of two new species: *Mastigonema stenellae* gen. et. sp. n. (Nematoda: Spiruroidea) and *Zalophotrema pacificum* sp. n. (Trematoda: Digenea). *Fishery Bulletin,* 71, 455 – 471 [68].

Dailey, M. D., and Stroud, R. (1978). Parasites and associated pathology observed in cetaceans stranded along the Oregon coast. *Journal of Wildlife Diseases,* 14, 503 – 511 [69].

Dailey, M. D., and Vogelbein, W. K. (1991). Parasite fauna of three species of Antarctic whales with reference to their use as potential stock indicators. *Fishery Bulletin,* 89, 355 – 365 [70].

Dailey, M. D., and Walker, W. A. (1978). Parasitism as a factor (?) in single strandings of Southern California cetaceans. *Journal of Parasitology,* 64, 593 – 596 [71].

Dans, S. L., Reyes, L. M., Pedraza, S. N., Raga, J. A., and Crespo, E. A. (1999). Gastrointestinal helminths of the Dusky dolphin, *Lagenorhynchus obscurus* (Gray, 1828), off Patagonia, in the Southwestern Atlantic. *Marine Mammal Science,* 15, 649 – 660 [72].

Davey, J. T. (1971). A revision of the genus *Anisakis* Dujardin, 1845 (Nematoda: Ascaridata). *Journal of Helminthology,* 45, 51 – 72 [73].

Dawes, B. (1946). *The Trematoda with special reference to British and other European forms*. Cambridge: University Press [74].

Delyamure, S. L. (1949). Towards the study of the helminth fauna of the dolphin, *Tursiops tursio*. *Fisheries Research Board of Canada,* 134, 1 – 5 [75].

Delyamure, S. L. (1955). *Helminthofauna of marine mammals (ecology and phylogeny)*. Moscow: Izdatel'stvo Akademii Nauk SSSR [76].

Delyamure, S. L. (1957). Dependence of the helminth fauna of dolphins of the Black and Azov seas on ecologico-geographical factors. *Fisheries Research Board of Canada,* 131, 1 – 10 [77].

Delyamure, S. L. (1961). About the necessity of exploring under all points of view and under the systematic point of view the parasites of industrial marine mammals. *Trudy Soveshchanil Ikhtiologicheskaya Komissiya Akademiya Nauk S.S.S.R.*, 21, 222 – 232 [78].

Delyamure, S. L. (1968). The find of *Diphyllobothrium stemmacephalum* (Cobbold, 1958) in the waters of the Soviet Union. *Parazitologiya (Leningr.)*, 4, 317 – 321 [79].

Delyamure, S. L. (1969). Enemies and parasites. In S. E. Kleinenberg, A. V. Yablokov, B. M. Bel'kobich and M. N. Tarasevich (Eds.), *Beluga (Delphinapterus leucas). Investigations of the species* (pp. 292 – 312). Jerusalem: Israel Program for Scientific Translations [80].

Delyamure, S. L. (1971). *Diphyllobothrium ponticum* n. sp. a new diphyllobothriid of the bottlenosed dolphin (*Tursiops truncatus* Montagu). *Sbornik Rabot po Gel'mintologii*, 123 – 125 [81].

Delyamure, S. L., and Kleinenberg, S. E. (1958). New data on the helminthfauna of the porpoise *Delphinapterus leucas*. *Biulleten Moskovskogo Obshchestva Ispytatelei Prirody. Otdel Biologicheskii*, 63, 25 – 32 [82].

Delyamure, S. L., and Skrjabin, A. S. (1966). *Diphyllobothrium polyrugosum* n. sp. parasite of *Orcinus orca* L. In S. L. Delyamure (Ed.), *Gel'mintofauna Zhivotnykh Yuzhnykh Morei* (pp. 3 – 9). Kiev: Naukova Dumka [83].

Delyamure, S. L., and Skrjabin, A. S. (1972). Helminths. In A. A. Berzin (Ed.), *The Sperm Whale* (pp. 278 – 304). Jerusalem: Israel Program for Scientific Translations [84].

Delyamure, S. L., and Skrjabin, A. S. (1986). Identification of the species *Diglonoporus balaenopterae* and *D. grandis* (Cestoda, Diphyllobothriidae). *Parazitologiya (Leningr.)*, 20, 69 – 72 [85].

Delyamure, S. L., Skrjabin, A. S., and Serdukov, A. M. (1985). Diphyllobothriata – flatworm parasites of man, mammals and birds. In T. I. Ponova and M. D. Sonc (Eds.), *Essentials of Cestodology* (pp. 1 – 200). Moscow: Nauka [86].

Demaree, R. S., Critchfield, R. L., and Tinling, S. P. (1997). *Oschmarinella macrorchis* sp.n. (Digenea: Campulidae) from the liver sinuses of a beaked whale, *Mesoplodon stejnegeri* (Cetacea: Ziphiidae). *Journal of Helminthology*, 64, 183 – 187 [87].

Desportes, G. (1985). *La nutrition des odontocètes en Atlantique Nord-Est (cotes Françaises- iles Feroe)*. Ph. D. Diss., University of Poitiers [88].

Dollfus, R. Ph. (1964). A propos de la recolte, a Banyuls d'un cystique de Cestode chez *Tursiops truncatus* (Montagu, 1821). Les cystiques de Cestodes chez les Cétacés et Pinnipedes. *Vie et Milieu,* 17, 177 – 204 [89].

Dollfus, R. Ph. (1965). Connait-on pluseiurs espèces de *Crassicauda* (Nematoda, Spiurata) chez les cétacés Ziphiidae? *Bulletin du Muséum National d'Histoire Naturelle, Paris,* 37, 865 – 878 [90].

Dollfus, R. Ph. (1966). Helminthofaune de *Kogia breviceps* (de Blainville, 1838) Cétacé Odontocète. *Annales de la Société des Sciences Naturelles de la Charente-Maritime,* 4, 3 – 6 [91].

Dollfus, R. Ph. (1973/74). *Pholeter* (Trematoda Digenea) from an intestinal cyst of *Stenella coeruleoalba* Meyen, 1833 (Odontoceti Delphinidae). Comments on the family Pholeteridae, R. Ph. Dollfus, 1939. List of helminths identified to date in *Stenella coeruleoalba* Meyen. *Investigations on Cetacea,* 5, 331 – 337 [92].

Domiciano, I. G., Domit, C., Broadhurst, M. K., Koch, M. S. and Bracarense, A. P. F. R. L. (2016) Assessing disease and mortality among small cetaceans stranded at a world heritage site in Southern Brazil. *Plos ONE,* 11, e0149295 [93].

Dougherty, E. C. (1943). Notes on the lungworms of porpoises and their occurrence on the California coast. *Proceedings of the Helmintological Society of Washington,* 10, 16 – 22 [94].

Ebert, M. B., and Schifino, V. A. L. (2013). New records of *Nasitrema attenuata* and *Nasitrema globicephalae* (Trematoda: Brachycladiidae) Neiland, Rice and Holden, 1970 in delphinids from South Atlantic. *Check List,* 9, 1538 – 1540 [95].

Edmonds, S. J. (1957). Acanthocephala. *BANZ Antarctic Research Expedition 1929-1931 Reports Series B, Zoology and Botany,* 6, 93 – 98 [96].

Edmonds, S. J. (1987). A note on the occurrence of *Bolbosoma capitatum* (Linstow, 1880) (Acanthocephala) from a false killer whale stranded on the coast of Western Australia. *Records of the Western Austalian Museum,* 13, 317 – 318 [97].

Fernández, M., Agustí, C., Aznar, F. J., and Raga, J. A. (2003a). Gastrointestinal helminths of Risso's dolphin *Grampus griseus* from the Western Mediterranean. *Diseases of Aquatic Organisms,* 55, 73 – 76 [98].

Fernández, M., Aznar, F. J., Latorre, A., and Raga, J. A. (1998a). Molecular phylogeny of the families Campulidae and Nasitrematidae (Trematoda) based on mtDNA sequence comparison. *International Journal for Parasitology,* 28, 767 – 775 [99].

Fernández, M., Aznar, F. J., Montero, F. E., Georgiev, B. B., and Raga, J. A. (2004). Gastrointestinal helminths of Cuvier's beaked whales, *Ziphius cavirostris*, from the western Mediterranean. *Journal of Parasitology,* 90, 418 – 420 [100].

Fernández, M., Balbuena, J. A., Pertusa, J. F., and Raga, J. A. (1995). Biometric varibaility of *Hadwenius tursionis* (Marchi, 1873) (Digenea, Campulidae) from the intestine of the bottlenose dolphin *Tursiops truncatus* (Montagu, 1821). *Systematic Parasitology,* 30, 67 – 76 [101].

Fernández, M., Balbuena, J. A., and Raga, J. A. (1994). *Hadwenius tursionis* (Marchi, 1873) n. comb. (Digenea, Campulidae) from the bottlenose dolphin *Tursiops truncatus* (Montagu, 1821) in the western Mediterranean. *Systematic Parasitology,* 28, 223 – 228 [102].

Fernández, M., Berón-Vera, B., García, N., Raga, J. A., and Crespo, E. A. (2003b). Food and parasites from two Hourglass dolphins, *Lagenorhynchus cruciger* (Quoy and Gaimard, 1824), from Patagonian waters. *Marine Mammal Science,* 19, 832 – 836 [103].

Fernández, M., Littlewood, D. T. J., Latorre, A., and Raga, J. A. (1998b). Phylogenetic relationships of the family Campulidae (Trematoda) based on 18S rRNA sequences. *Parasitology,* 117, 383 – 391 [104].

Figueroa, L., and Franjola, R. (1988). *Braunina cordiformis* Wolf 1903 (Digenea: Brauninidae) en delfín chileno *Cephalorhynchus eutropia* Gray, 1846 (Cetacea: Odontoceti). *Boletín Chileno de Parasitología,* 43, 71 – 72 [105].

Forrester, D. J., Odell, D. K., Thompson, N. P., and White, J. R. (1980). Morphometrics, parasites, and chlorinated hydrocarbon residues of pygmy killer whales from Florida. *Journal of Mammalogy,* 61, 356 – 360. [106].

Forrester, D. J., and Robertson, W. D. (1975). Helminths of rough-toothed dolphins, *Steno bredanensis* Lesson 1828, from Florida waters. *Journal of Parasitology,* 61, 922 [107].

Fraija-Fernández, N., Aznar, F. J., Raga, J. A., Gibson, D., and Fernández, M. (2014). A new brachycladiid species (Digenea) from Gervais' beaked whale *Mesoplodon europaeus* in north-western Atlantic waters. *Acta Parasitologica,* 59, 510 – 517 [108].

Fraija-Fernández, N., Olson, P. D., Crespo, E. A., Raga, J. A., Aznar, F. J., and Fernández, M. (2015). Independent host switching events by digenean parasites of cetaceans inferred from ribosomal DNA. *International Journal for Parasitology,* 45, 167 – 173 [109].

Fukui, T., and Morisita, T. (1939). Contributions to the knowledge of two species of Acanthocephala. *Volume Jubilare pro Prof. Sadao Yoshida,* 1, 137 – 139 [110].

Gallego, J., and Selva, J. M. (1979). *Skrjabinalius guevarai* n. sp. (Nematoda: Pseudaliidae), parásito pulmonar del delfín mular, *Tursiops truncatus* Montagu, 1821 (Cetacea: Delphinidae) en el Adriático. *Revista Ibérica de Parasitología,* 39, 203 – 208 [111].

Geraci, J. R., Dailey, M. D., and Aubin, D. J. (1978). Parasitic mastitis in the Atlantinc white-sided dolphin, *Lagenorhynchus acutus*, as a probable factor in herd productivity. *Journal of the Fisheries Research Board of Canada,* 35, 1350 – 1355 [112].

Gibson, D. I. (1973). Some 'blue' whale nematodes. *Parasitology,* 67, 7 – 8 [113].

Gibson, D. I., and Bray, R. A. (1997). *Oschmarinella albamarina* (Treschev, 1968) n. comb., a liver fluke from the killer whale *Orcinus orca* (L.) off the British coast. *Systematic Parasitology,* 36, 39 – 45 [114].

Gibson, D. I., and Harris, E. A. (1979). The helminth-parasites of cetaceans in the collection of the British Museum (Natural History). *Investigations on Cetacea,* 10, 309 – 324 [115].

Gibson, D. I., Harris, E. A., Bray, R. A., Jepson, P. D., Kuiken, T., Baker, J. R., and Simpson, V. R. (1998). A survey of the helminth parasites of cetaceans stranded on the coast of England and Wales during the period 1990-1994. *Journal of Zoology, London,* 244, 563 – 574 [116].

Golvan, Y. J. (1959). Acanthocèphales du genre *Coryonsoma* Lühe, 1904 Parasites de mammifères de'Alaska et de Midway. *Annales de Parasitologie Humaine Comparée,* 34, 288 – 321 [117].

Greenwood, A. G., Taylor, D. C., and Gauckler, A. (1979). Odontocete parasite – Some new host records. *Aquatic Mammalas,* 7, 23 – 25 [118].

Guiart, J. (1935). Cestodes parasites provenant des campagnes scientifiques de S.A.S. le Prince Albert ler de Monaco (1886-1913). *Résultat des Campagnes Scientifiques, Monaco,* 91, 1 – 100 [119].

Guimaraes, J. P., Febronio, A. M. B., Vergara-Parente, J. E., and Werneck, M. R. (2015). Lesions associated with *Halocercus brasiliensis* Lins de Almeida, 1933 in the lungs of dolphins stranded in the Northeast of Brazil. *Journal of Parasitology,* 101, 248 – 251 [120].

Hafeezullah, M. (1986). On a trematode (Digenea: Campulidae) of a marine mammal from Arabian Sea. *Records of the Zoological Survey of India,* 83, 41 – 48 [121].

Harada, J. (1931). Das Nervensystem von *Bolbosoma turbinella. Japanese Journal of Zoology,* 3, 161 – 199 [122].

Hatsushika, R., and Shirouzo, H. (1990). A new species of marine tapeworm, *Diphyllobothrium orcini* n. sp. (Cestoda: Pseudophyllidea) found from Killer Whale, *Orcinus orca* (Linnaeus, 1758) in Japan. *Japanese Journal of Parasitology,* 39, 566 – 573 [123].

Heckman, R. A., Jensen, L. A., Warnock, R. G., and Coleman, B. (1987). Parasites of the bowhead whale, *Balaena mysticetus. Great Basin Naturalist,* 47, 355 – 372 [124].

Herreras, M. V., Kaarstad, S. E., Balbuena, J. A., Kinze, C. C., and Raga, J. A. (1997). Helminth parasites of the digestive tract of the harbour porpoise *Phocoena phocoena* in Danish waters: a comparative geographical analysis. *Diseases of Aquatic Organisms,* 28, 163 – 167 [125].

Hoberg, E. P. (1990). *Trigonocotyle sexitesticulae* sp. nov. (Eucestoda: Tetrabothriidae): a parasite of pygmy killer whales (*Feresa attenuata*). *Canadian Journal of Zoology,* 68, 1835 - 1838 [126].

Hörning, B., and Pilleri, G. (1969). *Campula delphini* (Trematoda, Campulidae) bei *Delphinus delphis* and *Stenella styx* des westlichen Mittelmeeres. *Investigations on Cetacea,* 1, 193 – 196 [127].

Hutton, J., Blair, D., Slooten, E., and Dawson, S. M. (1987). Case studies of fluke-induced lesions in mesenteric lymph nodes of Hector's dolphins *Cephalorhynchus hectori. Diseases of Aquatic Organisms,* 2, 83 – 86 [128].

Iñiguez, A. M., Carvalho, V. L., Motta, M. R. A., Pinheiro, D. C. S. N., and Vicente, A. C. P. (2011). Genetic analysis of *Anisakis typica* (Nematoda: Anisakidae) from cetaceans of the northeast coast of Brazil: New data on its definitive hosts. *Veterinary Parasitology,* 178, 293 – 299 [129].

Iñiguez, A. M., Santos, C. P., and Vicente, A. C. P. (2009). Genetic characterization of *Anisakis typica* and *Anisakis physeteris* from marine mammals and fish from the Atlantic Ocean off Brazil. *Veterinary Parasitology,* 165, 350 – 356 [130].

Jabbar, A., Beveridge, I., Bryant and M. S. (2015) Morphological and molecular observations on the status of *Crassicauda magna*, a parasite of the subcutaneous tissues of the pygmy sperm whale, with a re-evaluation of the systematic relationships of the genus *Crassicauda. Parasitology Research*, 114, 835 – 841 [131].

Jefferson, T. A. (1988). *Phocoenoides dalli. Mammalian Species,* 319, 1 – 7 [132].

Jefferson, T. A., and Hung, S. K. (2004). *Neophocaena phocaenoides. Mammalian Species,* 746, 1 – 12 [133].

Johnston, T. H., and Best, E. W. (1937). Acanthocephala of the Australasian Antarctica expedition. *Scientific Reports Series C, Zoology and Botany/Australasian Antarctic Expedition,* 10, 1 – 20 [134].

Johnston, T. H., and Mawson, P. M. (1941). Nematodes from Australian marine mammals. *Records of the South Australian Museum,* 6, 429 – 434 [135].

Johnston, D. G., and Ridgway, S. H. (1969). Parasitism in some marine mammals. *Journal of the American Veterinary Medical Association,* 155, 1064 – 1072 [136].

Kagei, N. (1969). Life history of nematodes of the genus *Anisakis*. *Journal of the Fisheries Research Board of Canada (Translate Series),* 1324, 25 [137].

Kagei, N., Oshima, T., Kobayashi, A., Kumada, M., Koyama, T., Komiya, Y., Takemura, A., and Takemura, A. (1967). Survey of *Anisakis* spp. (Anisakinae, Nematoda) on marine mammals in the coast of Japan. *Japanese Journal of Parasitology,* 16, 427 – 435 [138].

Kagei, N., Tobayama, T., and Nagasaki, Y. (1976). On the helminthum of Franciscana, *Pontoporia blainvillei. Scientific Reports of the Whales Research Institute,* 28, 161 – 166 [139].

Kamo, H., Iwata, S., and Maejima, J. (1969). Re-examination of the diplogonadic tapeworm reported formerly from *Balaenoptera acutorostrata*. *Japanese Journal of Parasitology,* 18, 345 [140].

Kleinenberg, S. E., Yablokov, A. V., Belkovich, B. M., and Tarasevich, M. N. (1964). *Beluga (Delphinapterus leucas): Investigation of the species.* Moscow: Israel Program for Scientific Translations [141].

Kleinertz, S., Hermosilla, C., Ziltener, A., Kreicker, S., Hirzmann, J., Abdel-Ghaffar, F., and Taubert, A. (2014). Gastrointestinal parasites of free-living Indo-Pacific bottlenose dolphins (*Tursiops aduncus*) in the Northern Red Sea, Egypt. *Parasitology Research,* 113, 1405 – 1415 [142].

Klumov, S. K. (1963). Food and helminth fauna of whalebone whales (Mystacoceti) in the main whaling regions of the world oceans. *Journal of the Fisheries Research Board of Canada (Translate Series),* 71, 94 – 194 [143].

Krotov, A. I., and Delyamure, S. L. (1952). Some aspects of the parasitic worm fauna of mammals and birds of URSS. *Trudy Gel'mintologicheskoi Laboratorii Akademiya Nauk S.S.S.R.,* 6, 278 – 292 [144].

Kumar, V., Vercruysse, J., Kageruka, P., and Mortelmans, J. (1975). *Nasitrema attenuata* (Trematoda) infection of *Tursiops truncatus* and its potentialities as an aetiological agent of chronic pulmonary lesions. *Journal of Helminthology,* 49, 289 – 292 [145].

Kuramochi, T., Kikuchi, T., Okamura, H., Tatsukawa, T., Doi, H., Nakamura, K., Yamada, T. K., Koda, Y., Yoshida, Y., Matsuura, M., and Sakakibara, S. (2000). Parasitic helminth and epizoit fauna of finless porpoise in the Inland Sea of Japan and the Western North Pacific with a preliminary note on faunal difference by host's local population. *Memoirs of the National Science Museum, Tokyo,* 33, 83 – 95 [146].

Kuramochi, T., Wakao, H., Terasawa, F., Hagiwara S., Miyauchi, Y., Akamatsu, T., Yamada, T. K., Koda, Y., Sakai, T., Inaba, N., and Yoshida, Y. (2001). Stomach nematodes of the Family Anisakidae collected from the cetaceans stranded on or incidentally caught off the coasts of the Kanto Districts and adjoining areas. *Memoirs of the National Science Museum, Tokyo,* 37, 177 – 192 [147].

Lair, S., Measures, L. N. and Martineau, D. (2015) Pathologic findings and trends in mortality in the Beluga (*Delphinapterus leucas*) population of the St Lawrence Estuary, Quebec, Canada, from 1983 to 2012. *Veterinary Pathology,* 53, 22 – 36 [148].

Lambertsen, R. H. (1985). Taxonomy and distribution of a *Crassicauda* species (Nematoda: Spirurida) infecting the kidney of the common fin whale (*Balaenoptera physalus* Linné, 1758). *Journal of Parasitology,* 71, 485 – 488 [149].

Lamothe-Argumedo, R. (1987). Trematodos de mamiferos III. Hallazgo de *Synthesium tursionis* en *Phocoena sinus* en el Golfo de California Mexico. *Anales del Instituto de Biología, UNAM: serie Zoología,* 58, 11 – 20 [150].

Lane, E. P., de Wet, M., Thompson, P., Siebert, U., Wohlsein, P., and Plön, S. (2014). A systematic health assessment of Indian Ocean bottlenose (*Tursiops aduncus*) and Indo-Pacific humpback (*Sousa plumbea*). *Plos ONE,* 9, e107038 [151].

Layne, J. N., and Caldwell, D. K. (1964). Behavior of the Amazon dolphin, *Inia geoffrensis* (Blainville), in captivity. *Zoologica,* 49, 81 – 108 [152].

Lehnert, K., Raga, J. A., and Siebert, U. (2005). Macroparasites in stranded and bycaught harbour porpoises from German and Norwegian waters. *Diseases of Aquatic Organisms,* 64, 265 – 269 [153].

Lehnert, K., Samson-Himmelstjerna, G., Schaudien, D., Bleidorn, C., Wohlsein, P., and Siebert, U. (2010). Transmission of lungworms of harbour porpoises and harbour seals: Molecular tools determine potential vertebrate intermediate hosts. *International Journal for Parasitology,* 40, 845 – 853 [154].

Lehnert, K., Seibel, H., Hasselmeier, I., Wohlsein, P., Iversen, M., Nielsen, N. H., Heide-Jorgensen, M. P., Prenger-Berninghoff, E., and Siebert, U. (2014). Increase in parasite burden and associated pathology in harbour porpoises (*Phocoena phocoena*) in West Greenland. *Polar Biology,* 37, 321 – 331 [155].

Leiper, R. T., and Atkinson, E. L. (1914). Helminths of the British Antarctic expedition 1910-1913. *Proceedings of the Zoological Society of London,* 1, 222 – 226 [156].

Leonardi, M. S., Grandi, M. F., Garcia, N. A., Svendsen, G., Romero, M. A., González, R., and Crespo, E. A. (2011). A stranding of *Balaenoptera borealis* (Lesson 1828) from Patagonia, Argentina, with notes on parasite infestation and diet. *African Journal of Marine Science,* 33, 177 – 179 [157].

Lluch, J., Roca, V., and Navarro, P. (1987). *Campula palliata* (looss, 1885) Looss, 1901 (Digenea: Campulidae) parásito del delfín listado (*Stenella coeruleoalba* Mayen, 1833) (Mammalia: Delphinidae). *Real Sociedad Española de Historia Natural,* 83, 135 – 139 [158].

Loizaga de Castro, R., Leonardi, M. S., Grandi, M. F., García, N. A., and Crespo, E. A. (2011). Far from home: Record of a vagrant striped dolphin in Patagonia with notes on diet, parasites and age determination. *Mammalian Biology,* 76, 521 – 524 [159].

Loizaga de Castro, R., Vales, D. G., Degrati, M., García, N. A., Fernández, M., and Crespo, E. A. (2014). First record of cestode cysts of *Phyllobothrium delphini* (Phyllobothriidae) from dusky dolphins (*Lagenorhynchus obscurus*) off Argentine coast. *Hidrobiológica,* 24, 307 – 310 [160].

Lönnberg, E. (1892). Anatomische studien über skandinavische Cestoden, 2 (Zwei parasiten aus walfischen und zwei aus lamna cornubica). *Bihang til Kongliga Svenska Vetenskaps-akademiens Handlingar,* 24, 1 – 30 [161].

López-Neyra, C. R. (1958). Sobre unos vermes parasitos obtenidos en el *Ziphius cavirostris* Cuvier (Cetaceo Denticeto) varado en Almería en especial *Crassicauda giliakiana* (Spiruroidea; Tetrameridae). *Revista Ibérica de Parasitología,* 18, 13 – 18 [162].

Luque, J. L., Muniz-Pereira, L. C., Siciliano, S., Siqueira, L. R., Oliveira, M. S., and Vieira, F. M. (2010). Checklist of helminth parasites of cetaceans from Brazil. *Zootaxa,* 2548, 57 – 68 [163].

Lyster, L. L. (1940). Parasites of some Canadian sea mammals. *Canadian Journal of Research,* 18, 395 – 409 [164].

Mac Neill, A. C., Neufeld, J. L., and Webster, W. A. (1975). Pulmonary nematodiasis in a narwhale. *The Canadian Veterinary Journal,* 16, 52 – 55 [165].

Magalhaes-Pinto P. R., Muniz-Pereira, L. C., Alves, V. C., and Siciliano, S. (2004). First report of a helminth infection for Bryde's whale, *Balaenoptera edeni* Anderson, 1878 (Cetacea, Balaenopteridae). *Latin American Journal of Aquatic Mammals,* 3, 167 – 170 [166].

Malatesta, T., Frati, R., Cerioni, S., Agrimi, U., and Di Guardo, G. (1998). *Ogmogaster antarcticus* Johnston, 1931 (Digenea: Notocotylidae) in *Balaenoptera physalus* (L.): first record in the Mediterranean Sea. *Systematic Parasitology,* 40, 63 – 66 [167].

Manfredi, M. T., Cammarata, G., Dini W., Ganduglia, S., and Repetto, G. (1994). Infestazione da *Pholeter gastrophilus* in Delphinidae della specie *Stenella coeruleoalba* nel Mar Ligure. *PRAXIS Veterinaria,* 15, 27 – 28 [168].

Manfredi, M. T., Dini, W., Ganduglia, S., Podesta, M., and Repetto, G. (1992). Parasitological findings in striped dolphins *Stenella coeruleoalba. European Research on Cetaceans,* 6, 218 – 219 [169].

Margolis, L., and Arai, H. P. (1989). Parasites of marine mammals. In M. J. Kennedy (Ed.), *Synopsis of the Parasites of Vertebrates of Canada* (pp. 1 – 26). Edmonton: Alberta Agriculture Publication [170].

Margolis, L., and Dailey, M. D. (1972). Revised annotated list of parasites from sea mammals caught off the west coast of North America. *NOAA Technical Report,* 647, 1 – 23 [171].

Margolis, L., and Pike, G. C. (1955). Some helminth parasites of Canadian pacific whales. *Journal of the Fisheries Research Board of Canada,* 12, 97 – 120 [172].

Marigo, J., Cunha, H. A., Bertozzi, C. P., Souza, S. P., Rosas, F. C. W., Cremer, M. J., Barreto, A. S., Oliveira, L. R., Capozzo, H. L., Valente, A. L. S., Santos, C. P., and Vicente, A. C. P. (2015). Genetic diversity and population structure of *Synthesium pontoporiae* (Digenea, Brachycladiidae) linked to its definitive host stocks, the endangered Franciscana dolphin, *Pontoporia blainvillei* (Pontoporiidae) off the coast of Brazil and Argentina. *Journal of Helminthology,* 89, 19 – 27 [173].

Marigo, J., Paulo A. C., Schifino, A. L., Measures, L., and Portes C. (2008). Redescription of *Synthesium pontoporiae* n. comb. with notes on *S.*

tursionis and *S. seymouri* n. comb. (Digenea: Brachycladiidae Odhner, 1905). *Journal of Parasitology,* 94, 505 – 514 [174].

Marigo, J., Rosas, F. C. W., Andrade, A. L. V., Oliveira, M. R., Dias, R. A., and Catao-Dias, J. L. (2002). Parasites of franciscana (*Pontoporia blainvillei*) from São Paulo and Paraná States, Brazil. *Latin American Journal of Aquatic Mammals,* 1, 115 – 122 [175].

Marigo, J., Ruoppolo, V., Rosas, F. C. W., Valente, A. L. S., Oliveira, M. R., Dias, R. A., and Catao-Dias, J. L. (2010). Helminths of *Sotalia guianensis* (Cetacea: Delphinidae) from the South and Southeastern Coasts of Brazil. *Journal of Wildlife Diseases,* 46, 599 – 602 [176].

Markowski, S. (1955). Cestodes of whales and dolphins from the discovery collections. *Discovery Reports,* 27, 377 – 395 [177].

Mateu, P., Raga, J. A., and Aznar, F. J. (2011). Host specificity of *Oschmarinella rochebruni* and *Brachycladium atlanticum* (Digenea: Brachycladiidae) in five cetacean species from western Mediterranean waters. *Journal of Helminthology, 85*, 12 – 19 [178].

Mateu, P., Raga, J. A., Fernandez, M., and Aznar, F. J. (2014). Intestinal helminth fauna of striped dolphins (*Stenella coeruleoalba*) in the western Mediterranean: No effects of host body length, age and sex. *Marine Mammal Science,* 30, 961 – 977 [179].

Mattiucci, S., Cipriani, P., Webb, S. C., Paoletti, M., Marcer, F., Bellisario, B., Gibson, D. I., and Nascetti, G. (2014). Genetic and morphological approaches distringuish three sibling species of the *Anisakis simplex* species complex, with a species designation as *Anisakis berlandi* n. sp. for *A. simplex* sp. *C* (Nematoda: Anisakidae). *Journal of Parasitology,* 100, 199 – 214 [180].

Mattiucci, S., and Nascetti, G. (2008). Advances and trends in the moleculars systematics of anisakid nematodes, with Implications for their evolutionary, ecology and host–parasite co-evolutionary processes. *Advances in Parasitology,* 66, 47 – 148 [181].

Mattiucci, S., Nascetti, G., Cianchi, R., Paggi, L., Arduino, P., Margolis, L., Brattey, J., Webb, S., D'Amelio, S., Orecchia, P., and Bullini, L. (1997). Genetic and ecological data on the *Anisakis simplex* complex, with evidence for a new species (Nematoda, Ascaridoidea, Anisakidae). *Journal of Parasitology,* 83, 401 – 416 [182].

Mattiucci, S., Nascetti, G., Dailey, M., Webb S. C., Barros, N. B., Cianchi, R., and Bullini, L. (2005). Evidence for a new species of *Anisakis* Dujardin, 1845: morphological description and genetic relationships between congeners (Nematoda: Anisakidae). *Systematic Parasitology,* 61, 157 – 171 [183].

Mattiucci, S., Paggi, L., Nascetti, G., Santos, C. P., Costa, G., Di Beneditto, A. P., Ramos, R., Argyrou, M., Cianchi, R., and Bullini, L. (2002). Genetic markers in the study of *Anisakis typica* (Diesing, 1860): larval identification and genetic relationships with other species of *Anisakis* Dujardin, 1845 (Nematoda: Anisakidae). *Systematic Parasitology,* 51, 159 – 170 [184].

Mattiucci, S., Paoletti, M., Webb, S., D'Amelio, S., Orecchia, P., and Bullini, L. (2009). *Anisakis nascettii* n. sp. (Nematoda: Anisakidae) from beaked whales of the southern hemisphere: morphological description, genetic relationships between congeners and ecological data. *Systematic Parasitology,* 74, 199 – 217 [185].

McColl, K. A., and Obendorf, D. L. (1982). Helminth parasites and associated pathology in stranded Fraser's dolphins, *Lagenodelphis hosei* (Fraser, 1956). *Aquatic Mammals,* 9, 30 – 34 [186].

McIntosh, A. (1960). A new campulid trematode *Hunterotrema caballeroi* n. g., n. sp. from an amazon dolphin *Inia geoffrensis*. *Libro homenaje al Dr. Eduardo Caballero y Caballero 1930-1960. Instituto Politécnico Nacional de Mexico,* 207 – 208 [187].

McKenzie, J., and Blair, D. (1983). Parasites from Hector's dolphin (*Cephalorhynchus hectori*). *New Zealand Journal of Zoology,* 10, 126 – 127 [188].

Measures, L. N. (1993). Annotated list of metazoan parasites reported from the blue whale, *Balaenoptera musculus*. *Journal of the Helminthological Society of Washington,* 60, 62 – 66 [189].

Measures, L. N. (2001). Lungworms of marine mammals. In W. M. Samuel, M. J. Pybus and A. A. Kocan (Eds.), *Parasitic diseases of wild mammals* (pp. 279 – 300). Iowa: Iowa State University Press [190].

Measures, L. N., Béland, P., Martineau, D., and De Guise, S. (1995). Helminths of an endangered population of belugas, *Delphinapterus leucas*, in the St. Lawrence estuary, Canada. *Canadian Journal of Zoology,* 73, 1402 – 1409 [191].

Melo, O. P., Ramos, R. M. A., and Di Beneditto, A. P. M. (2006). Helminths of the marine tucuxi, *Sotalia fluviatilis* (Gervais, 1853) (Cetacea: Delphinidae), in northern Rio de Janeiro State, Brazil. *Brazilian Archives of Biology and Technology,* 49, 145 – 148 [192].

Migaki, G., Van Dyke, D., and Hubbard, R. C. (1971). Some hitopathological lesions caused by helminths in marine mammals. *Journal of Wildlife Diseases,* 7, 281 – 289 [193].

Mignucci-Giannoni, A., Hoberg, E. P., Siegel-Causey, D., and Williams, E. H. (1998). Metazoan parasites and other symbionts of cetaceans in the Caribbean. *Journal of Parasitology,* 84, 939 – 946 [194].

Miyazaki, N., Fujise, Y., and Iwata, K. (1998). Biological analysis of a mass stranding of melon-headed whales (*Peponocephala electra*) at Aoshima, Japan. *Bulletin of the National Science Museum, Tokyo,* 24, 31 – 60 [195].

Morales-Vela, B., and Olivera-Gomez, L. D. (1993). Varamiento de calderones *Globicephala macrorhynchus* (Cetacea: Delphinidae) en la Isla de Cozumel, Quintana Roo, México. *Anales del Instituto de Biología, UNAM: serie Zoología,* 64, 177 – 180 [196].

Morimitsu, T., Kawano, H., Torihara, K., Kato, E., and Koono, M. (1994). Histopathology of eighth cranial nerve of mass stranded dolphins at Goto islands, Japan. *Journal of Wildlife Diseases,* 28, 656 – 658 [197].

Morimitsu, T., Nagai, T., Ide, M., Kawano, H., Naichuu, A., Koono, M., and Ishii, A. (1987). Mass stranding of odonceti caused by parasitogenic eighth cranial neuropathy. *Journal of Wildlife Diseases,* 23, 586 – 590 [198].

Moser, M., and Rhinehart, H. (1993). The lungworm, *Halocercus* spp. (Nematoda: Pseudaliidae) in cetaceans from California. *Journal of Wildlife Diseases,* 29, 507 – 508 [199].

Muniz-Pereira, L. C., Vicente, J. J., and Noronha, D. (1999). Helminth parasites of whales in Brazil. *Revista Brasileira de Zoologia,* 16, 249 – 252 [200].

Muñoz, G., and Olmos, V. (2008). Revisión bibliográfica de especies endoparásitas y hospedadoras de sistemas acuáticos de Chile. *Revista de Biología Marina y Oceanografía,* 43, 173 – 245 [201].

Murav'eva, S. I., and Treskchev, V. V. (1970). New cestoda *Priapocephalus eschrichtii sp.* n. (Tetrabothriidae) - a parasite of the gray whale (*Eschrichtius gibbosus* Erzleben, 1777) from the Chuhotsk Sea. *Vestnik Zoologichesky Akademiia Nauk S.S.S.R.,* 84 – 86 [202].

Neiland, K. A. (1962). Alaskan species of acanthocephalan genus *Corynosoma* Luehe, 1904. *Journal of Parasitology,* 48, 69 – 75 [203].

Neiland, K. A., Rice, D. W., and Holde, B. L. (1970). Helminth of marine mammals I, the genus *Nasitrema*, air sinus flukes of delphinid Cetacea. *Journal of Parasitology,* 56, 305 – 316. [204].

Nikolov, P. N., Cappozzo, H. L., Berón-Vera, B., Crespo, E. A., Raga J. A., and Fernández, M. (2010). Cestodes from Hector's beaked whale (*Mesoplodon hectori*) and spectacled porpoise (*Phocoena dioptrica*) from Argentinean waters. *Journal of Parasitology,* 96, 746 – 751 [205].

Odhner, T. (1905). *Die Trematoden de Arktischen Gebietes.* Leipzing: G. Fischer [206].

Oliveira, J. B., Morales, J. A., González-Barrientos, R. C., Hernández-Gamboa, J., and Hernández-Mora, G. (2011). Parasites of cetaceans stranded on the Pacific coast of Costa Rica. *Veterinary Parasitology,* 182, 319 – 328 [207].

Olson, P. D., and Caira, J. (1999). Evolution of the major lineages of tapeworms (Platyhelminthes: Cestoidea) inferred from 18S ribosomal DNA and elongation factor-1alpha. *Journal of Parasitology,* 85, 1134 – 1159 [208].

Oshima, T. (1972). *Anisakis* and anisakiasis in Japan and adjacent area. In K. Morishita, Y. Komiya, and H. Matsubayashi (Eds.), *Progress of Medical Parasitology in Japan* (pp. 305 – 393). Tokyo: Meguro Parasitological Museum [209].

Ozaki, Y. (1935). Trematode parasites of Indian porpoise *Neophocaena phocaenoides* Gray. *Journal of Science of the Hiroshima University,* 3, 115 – 138 [210].

Palsson, J., and Sigurjonsson, J. (1990). Parasitic nematodes from stomach of minke whales (*Balaenoptera acutorostrata*) off Iceland. *Reports of the International Whaling Commission,* SC42, 4 [211].

Parsons, E. C. M., Overstreet, R. M., and Jefferson, T. A. (2001). Parasites from Indo-Pacific hump-backed dolphins (*Sousa chinensis*) and finless porpoises (*Neophocaena phocaenoides*) stranded in Hong Kong. *Veterinary Record,* 23, 776 – 780 [212].

Pearson, J. C., and Courtney, C. H. (1977). *Pholeter anterouterus* Fischthal and Nasir, 1974 (Digenea: Opisthorchiidae) redescribed, together with remarks on the genera *Pholeter* Odhner, 1914 and *Phocitrema* Goto and Ozaki, 1930 and their relationship to the centrocestine heterophyids. *Parasitology,* 74, 255 – 271 [213].

Pekmezci, G. Z., Yardimci, B., Gürler, A. T., Bölükbas, C. S., Açici, M. and Umur, S. (2013) Survey on the presence of nematodes and associated with pathology in marine mammals from turkish waters. *Kafkas Üniversitesi Veteriner Fakültesi Dergisi,* 19, 1035 – 1038 [214].

Pendergraph, G. E. (1971). First report of the acanthocephalan *Bolbosoma vasculosum* (Rudolphi, 1819) from the pygmy sperm whale, *Kogia breviceps*. *Journal of Parasitology,* 57, 1109 [215].

Petter, A. J. (1971). Redescription of *Contracaecum lobulatum* (Schneider, 1866) (Ascaridoidea), a parasite of two species of *Platanista* (Cetacea). *Investigations on Cetacea,* 3, 53 – 58 [216].

Pinedo, M. C., Barreto, A. S., Lammardo, M. P., Andrade, A. L. V., and Geracitano, L. (2002). Northernmost records of the spectacled porpoise, Layard's beaked whale, Commerson`s dolphin and Peales's dolphin in the southwestern Atlantic Ocean. *Aquatic Mammals,* 28, 32 – 37 [217].

Popova, T. I., Mozgovoi, A. A., and Korotova, E. M. (1971). Helminth fauna of marine mammals of the White Sea. *Sbornik Rabot po Gel'mintologii,* 90, 306 – 312 [218].

Price, E. W. (1932). The trematode parasites of marine mammals. *Proceedings of the United States National Museum,* 81, 1 – 68 [219].

Prudhoe, S. (1969). Cestodes from fish, birds and whales. *BANZ Antarctic Research Expedition 1929-1931, Reports Series B, Zoology and Botany,* 8, 171 – 193 [220].

Quiazon, K. M. A., Santos, M. D. and Yoshinaga, T. (2013) *Anisakis* species (Nematoda: Anisakidae) of dwarf sperm whale *Kogia sima* (Owen, 1866) stranded off the Pacific coast of southern Philippine archipelago. *Veterinary Parasitology*, 197, 221 – 230 [221].

Quiñones, R., Giovannini, A., Raga, J. A., and Fernández, M. (2013). Intestinal helminth fauna of bottlenose dolphin *Tursiops truncatus* and common dolphin *Delphinus delphis* from the Western Mediterranean. *Journal of Parasitology,* 99, 576 – 579 [222].

Raga, J. A. (1985). Découverte d'un cestode parasite encore non signalé chez *Lagenodelphis hosei* (Fraser, 1956) (Cetacea, Delphinidae). *Bulletin du Muséum National d'Histoire Naturelle, Paris,* 7, 415 – 418 [223].

Raga, J. A. (1986). *Oschmarinella mascomai* n. sp. (Trematoda: Campulidae) parasite des canaux billiaires de *Stenella coeruleoalba* (Meyen, 1833) (Cetacea: Delphinidae) en Méditerranée. *Annales de Parasitologie Humaine Comparée,* 61, 633 – 636 [224].

Raga, J. A. (1987). Redescription de *Crassicauda grampicola* Johnston and Mawson, 1941 (Nematoda: Spirurida), parasite de *Grampus griseus* (Cuvier, 1812) (Cetacea: Delphinidae). *Vie et Milieu,* 37, 215 – 219 [225].

Raga, J. A. (1994). Parasitismus bei den Cetacea. In D. Robineau, R. Duguy, and M. Klima (Eds.), *Handbuch der Säugetiere Europas. Teil IA: Wale und Delphine I* (pp. 132 – 179). Wiesbaden: AULA-Verlag [226].

Raga, J. A., Aguilar, A., Fernández, J. P., and Carbonell, E. (1987a). Parasitofauna de *Balaenoptera physalus* (L., 1758) (Cetacea. Balaenopteridae) en las costas Atlánticas españolas. I. Sobre la aparición de *Ogmogaster antarcticus* Johnston, 1931 (Trematoda: Notocotylidae). *Revista Ibérica de Parasitología,* 46, 237 – 241 [227].

Raga, J. A., Aznar, F. J., Balbuena, J. A., and Dailey, M.D. (1994). *Hadwenius pontoporiae* sp. n. (Digenea: Campulidae) from the Intestine of Franciscana (Cetacea: Pontoporiidae) in Argentinian Waters. *Journal of the Helminthological Society of Washington,* 61, 45 – 59 [228].

Raga, J. A., and Balbuena, J. A. (1987). Algunas carcateristicas zoogeograficas de los helmintos de los cetaceos en el Mediterráneo con especial referencia a la helmintofauna del delfin listado. In V. Sans-Coma, S. Mas-Coma, and J. Gosálbes (Eds.), *Mamíferos y Helmintos* (pp. 195 – 201). Barcelona: Ketres Editora [229].

Raga, J. A., and Balbuena, J. A. (1988). *Leucasiella delamurei* sp. n. (Trematoda: Campulidae), a parasite of *Globicephala melaena* (Trail, 1809) (Cetacea: Delphinidae) in the Mediterranean Sea. *Helminthologia,* 25, 95 – 102 [230].

Raga, J. A., and Balbuena, J. A. (1990). A new species of the genus *Crassicauda* Leiper and Atkinson, 1914 (Nematoda: Spiruroidea) from the penis of *Globicephala melas* (Traill, 1809) (Cetacea: Globicephalidae) in the western Mediterranean Sea. *Annales de Parasitologie Humaine Comparée,* 65, 255 – 261 [231].

Raga, J. A., and Balbuena, J. A. (1993). Parasites of the long-finned pilot whale, *Globicephala melas* (Traill, 1809), in European waters. *Report of the International Whaling Commission,* 14, 391 – 406 [232].

Raga, J. A., and Carbonell, E. (1985). New data about parasites on *Stenella coeruleoalba* (Meyen, 1833) (Cetacea: Delphinidae) in the western Mediterranean Sea. *Investigations on Cetacea,* 17, 207 – 213 [233].

Raga, J. A., Carbonell, E., Raduán, M. A., and Blanco, C. (1985). Sobre la presencia de *Pholeter gastrophilus* (Kossack, 1910) (Trematoda: Troglotrematidae), en quistes estomacales de *Tursiops truncatus* y *Stenella coeruleoalba* (Cetacea: Delphinidae), en las costas españolas del Mediterráneo. *Revista Ibérica de Parasitología,* 45, 123 – 128 [234].

Raga, J. A., Casinos, A., Filella, S., and Raduán, M. A. (1982). Notes on cetaceans of the Iberian coasts. V. *Crassicauda grampicola* Johnston and Mawson, 1941, (Nematoda) cause of injuries in the pterygoids of some specimens of *Grampus griseus*. *Säugetierkundliche Mitteilungen,* 30, 315 – 318 [235].

Raga, J. A., Fernández, M., Balbuena, J. A., and Aznar, F. J. (2009). Parasites In W. F. Perrin, B. Würsig, J. G. M. Thewissen (Eds.), *Encyclopedia of Marine Mammals* (2nd. Edition, pp. 821 – 830). San Diego: Academic Press.

Raga, J. A., Kinze, C. C., Balbuena, J. A., Ortiz, T., and Fernández, M. (1989). New data on helminth parasites of the harbour porpoise *Phocoena phocoena* in Danish waters. Paper presented at the *Third Annual Conference of the European Cetacean Society*, La Rochelle, France, February 24 - 26 [236].

Raga, J. A., Petter, A., and Duguy, R. (1987b). Catalogue des parasites de cétacés des collections du Museé Océanographique de La Rochelle. *Bulletin du Muséum National d'Histoire Naturelle, Paris,* 119, 159 – 168 [237].

Rausch, R. L. (1964). Studies on the helminth fauna of Alaska. XLI. Observations on cestodes of the genus *Diplogonoporus* Lönnberg, 1892 (Diphyllobothriidae). *Canadian Journal of Zoology,* 42, 1049 – 1069 [238].

Rausch, R. L., and Fay, F. H. (1966). Studies on the helminth fauna of Alaska XLIV. Revision of *Ogmogaster* Jaggerskïod, 1891, with a description of *O. pentalineatus* sp. n. (Trematoda: Notocotylidae). *Journal of Parasitology,* 52, 26 – 38 [239].

Rausch, R. L., and Hilliard, D. K. (1970). Studies on the helminth fauna of Alaska. XLIX. The occurrence of *Diphyllobothrium latum* in Alaska with notes in other species. *Canadian Journal of Zoology,* 48, 1201 – 1219 [240].

Rausch, R. L., and Margolis, L. (1969). *Plicobothrium globicephalae* gen. et sp. nov. (Cestoda: Diphyllobothriidae) from the pilot whale, *Globicephala melaena* Traill, in Newfoundland Waters. *Canadian Journal of Zoology,* 47, 745 – 750 [241].

Rausch, R. L., and Rice, D. W. (1970). *Ogmogaster trilineatus* sp. n. (Trematoda: Notocotylidae) from the fin whale, *Balaenoptera physalus* L. *Proceedings of the Helminthological Society of Washington,* 37, 196 – 200 [242].

Rees, G. (1953). A record of some parasitic worms from whales in the Ross Sea. *Parasitology,* 43, 27 – 34 [243].

Reyes, J. C., and Van Waerebeek, K. (1993). Aspects of the biology of Burmeister's porpoise from Peru. *Reports of the International Whaling Commission,* 16, 349 – 364 [244].

Rice, D. W. (1963). Progress report on biological studies of the larger cetacean in the waters off California. *Norsk Hvalfangst-tid,* 7, 181 – 187 [245].

Rice, D. W., and Wolman, A. A. (1971). The life history and ecology of the gray whale (*Eschrichtius robustus*). *American Society of Mammalogists, Special Publication,* 3, 1 – 142 [246].

Ridgway, S. H., and Dailey, M. D. (1972). Cerebral and cerebellar involvement of trematode parasites in dolphins and their possible role in stranding. *Journal of Wildlife Diseases,* 8, 33 – 43 [247].

Rogan, E., Baker, J. R., Jepson, P. D., Berrow, S., and Kiely, O. (1997). A mass stranding of white-sided dolphins (*Lagenorhynchus acutus*) in Ireland: biological and pathological studies. *Journal of Zoology, London,* 242, 217 – 227 [248].

Rokicki, J., Berland, B., and Wróblewski, J. (1997). Helminths of the harbour porpoise, *Phocoena phocoena* (L.), in the southern Baltic. *Acta Parasitologica,* 42, 36 – 39 [249].

Romero, M. A., Fernández, M., Dans, S. L., García, N. A., González, R., and Crespo, E. A. (2014). Gastrointestinal parasites of bottlenose dolphins *Tursiops truncatus* from the extreme southwestern Atlantic, with notes on diet composition. *Diseases of Aquatic Organisms,* 108, 61 – 70 [250].

Rosas, F. C. W., Monteiro-Filho, E. L. A., Marigo, J., Santos, A., Andrade, A. L. V., Rautenberg, M., Oliveira, M. R., and Bordignon, M. O. (2002). The striped dolphin, *Stenella coeruleoalba* (Cetacea: Delphinidae), on the coast of Sao Paulo State, southeastern Brazil. *Aquatic Mammals,* 28, 60 – 66 [251].

Ross, G. J. B., and Bass, A. J. (1971). Shark attack on an ailing dolphin *Stenella coerueoalba* (Meyen). *South African Journal of Science,* 67, 413 – 414 [252].

Santos, C. P., Rohde, K., Ramos, R., and Di Beneditto, A. P. (1996). Helminths of cetaceans on the Southeastern coast of Brazil. *Journal of the Helminthological Society of Washington,* 63, 149 – 152 [253].

Schmidt, G. D., and Dailey, M. D. (1971). Zoogeography and the generic status of *Polymorphus* (*Polymorphus*) *cetaceum* (Johnston et Best, 1942) comb. n. (Acanthocephala). *Proceedings of the Helminthological Society of Washington,* 38, 137 [254].

Schryver, H. F., Medway, W., and Williams, J. F. (1967). The stomach fluke *Braunina cordiformis* in the Atlantic bottlenose dolphin. *Journal of the American Veterinary Medical Association,* 151, 884 – 886 [255].

Scott, D. M., and Fisher, H. D. (1958). Incidence of a parasitic ascarid, *Porrocaecum decipiens*, in the common porpoise, *Phocoena phocoena*, from the lower Bay of Fundy. *Journal of the Fisheries Research Board of Canada,* 15, 1 – 4 [256].

Shirouzo, H., Hatsushira, R., and Okino, T. (1999). Morphological studies on the diphyllobothriid tapeworms from killer whale, *Orcinus orca* (Linnaeus, 1758) and bottlenose dolphin, *Tursiops truncatus* (Montague, 1821). *Japanese Journal of Parasitology, Tokyo,* 39, 566 – 573 [257].

Shults, L. M. (1979). *Ogmogaster antarcticus* Johnston, 1931 (Trematoda: Notocotylidae) from the bowhead whale, *Balaena mysticetus* L., at Barrow Alaska. *Canadian Journal of Zoology,* 57, 1367 – 1368 [258].

Shults, L. M., Fay, F. H., and Hall, J. D. (1982). Helminths from Stejneger's beaked whale *Mesoplodon stejnegeri* and Risso's dolphin *Grampus griseus* in Alaska. *Proceedings of the Helminthological Society of Washington,* 49, 146 – 147 [259].

Skrjabin, A. S. (1947). *Oschmarinella sobolevi* n. gen. n. sp. Novaya trematoda iz pecheni kita. *Doklady Akademiia Nauk,* 57, 857 – 859 [260].

Skrjabin, A. S. (1959). New helminth species of marine mammals in the Pacific Ocean and Dalbnevostochnij Sea. *Izvestiya Crimskogo Pedagogicheskogo Instituta,* 34, 99 – 118 [261].

Skrjabin, A. S. (1960). Helminth fauna of marine mammals in the Pacific Ocean and Far-East Sea. *Trudy perboi Nauchnoi konferensii Prepodabatee Biologicheskij, selbskojosyaisthenij y Jimicheskij Dissiplin. Pedag. Inst. S.S.S.R.,* 15, 225 – 241. [262].

Skrjabin, A. S. (1965). Larval forms of helminths from marine mammals of far Eastern seas. *Fisheries Research Board of Canada Translation Series,* 827, 9. [263].

Skrjabin, A. S. (1967). A gigantic diphyllobothriid *Polygonoporus giganticus* n. g., n. sp. - sperm whale parasite. *Parazitologiya (Leningr.),* 1, 131 – 135 [264].

Skrjabin, A. S. (1969). A new species of trematodes of the genus *Ogmogaster* (Notocotylidae) parasite of whalebone whales. *Zoologichesky Zhurnal,* 48, 1882 – 1885 [265].

Skrjabin, A. S. (1970a). A new species of acanthocephalans, *Bolbosoma tuberculata* sp. n. (Fam. Polymorphidae Meyer, 1931) - a parasite of whales. *Parazitologiya (Leningr),* 4, 334 – 337 [266].

Skrjabin, A. S. (1970b). A new trematode *Cetitrema foliformis* g. et. sp. n., a parasite of the liver of Bryde's whale. *Parazitologiya (Leningr),* 4, 34 – 38 [267].

Skrjabin, A. S. (1970c). A new trematode *Lecithodesmus petrowi* n. sp. parasite of the Bryde's whale. *Naucnye Doklady Vyssej Skoly Biologiceskie Nauki,* 7, 7 – 11 [268].

Skrjabin, A. S. (1970d). The first finding of larvae of the cestode *Monorygma delphini* (Tetraphyllidea, Phyllobothriidae) in Antarctic whales. *Zoologichesky Zhurnal,* 49, 1721 – 1722 [269].

Skrjabin, A. S. (1972). Larvae of cestodes of the genus *Phyllobothrium* Beneden, 1850 (Tetraphyllidea), parasites of whales and other marine animals. *Parazitologiya,* 6, 426 – 434 [270].

Skrjabin, A. S. (1973). New data in the study of the species *Crassicauda crassicauda* (Nematoda, Spirurida). *Biologicheskie Nauki,* 16, 7 – 12 [271].

Skrjabin, A. S. (1974). First finding of the trematodes *Lecithodesmus goliath* (Van Beneden, 1858) Braun, 1902 (Familia Campulidae Odhner, 1926) in the Antarctic whales. *Izvestiya Tijookeanskogo Naukhoissledovatel'skogo Instituta R'lkogo Jozyaitsua i Okeanografii (Tinro),* 88, 139 – 144 [272].

Skrjabin, A. S. (1978). The dependence of the helminth fauna of sperm whales on their sex, geographic distribution and life-style. *First All-Union Meeting of parasitocoenologists,* 3, 146 – 148 [273].

Skrjabin, A. S., and Delyamure, S. L. (1982). Morphology, systematics, and geographical distribution of nematodes – commensals of whalebone whales. In (Ed.), *Parazity i parazitozy cheloveka i zhivothnykh* (pp. 166 – 170). Kiev: Sbornik nauchnykh trudov [274].

Skrjabin, A. S., Delyamure, S. L., and Alexeiev, E. V. (1963). First finding of the strigeid *Braunina cordiformis* Wolf, 1903, (Suborder Cyathocotylata Sudarilov, 1959) in the Black Sea dolphin. *Gel'minths cheloveka, zhibotnij i borba c'nimi, Izd. Ad. S. S. S. R.,* 119 – 121 [275].

Skrjabin, A. S., and Murav'eva S. I. (1971). New cestode, *Tetrabothrius egregius* n. sp., parasite of *Balaenoptera physalus*. *Nauchnye Doklady Vysshei Shkoly, Biologicheskie Nauki,* 7, 17 – 21 [276].

Skrjabin, A. S., and Murav'eva S. I. (1972). First finding of *Tetrabothrius curilensis* (Tetraphyllidea) in *Physeter catodon* L., caught in the Antarctic waters. *Vestnik Zoologii,* 6, 85 – 87 [277].

Spratt, D. (2002). Parasites and pathology of the respiratory tracts of native and feral mammals in Australia - A review. *Australian Mammalogy,* 24, 177 – 192 [278].

Tajima, Y., Maeda, K., and Yamada, T. K. (2015). Pathological findings and probable causes of the death of Stejneger's beaked whales (*Mesoplodon stejnegeri*) stranded in Japan from 1999 and 2011. *Journal of Veterinary Medicine,* 77, 45 – 51 [279].

Tantalean, M., and Cabrera, R. (1999). Algunos helmintos de la marsopa espinosa, *Phocoena spinipinnis* de la Reserva Nacional de Paracas, Perú. *Parasitología al día,* 23, 56 – 57 [280].

Testa, J., and Dailey, M. D. (1977). Five new morphotypes of *Phyllobothrium delphini* (Cestoda: Tetraphyllidea), their relationship to existing morphotypes, and their zoogeography. *Bulletin of the Southern California Academy of Sciences,* 76, 99 – 110 [281].

Testi, F., and Pilleri, G. (1968). Polmonite verminosa da nematodi nel delfino (*Delphinus delphis* L.). *Acta Medica Veterinaria,* 14, 523 – 532 [282].

Testi, F., and Pilleri, G. (1969). Verminous pulmonitis induced by Nematoda (*Halocercus,* Pseudaliidae) in the dolphin (*Delphinus delphis* L.). *Investigations on Cetacea,* 1, 179 – 188 [283].

Tobayama, T., Nishiwaki, M., and Yang, H. C. (1973). Records of the Fraser's Sarawak dolphin (*Lagenodelphis hosei*) in the western North Pacific. *The Scientific Reports of the Whales Research Institute,* 25, 251 – 263 [284].

Tomilin, A. G. (1967). Cetacea. In *Mammals of the U.S.S.R., and adjacent countries* (pp. 1 – 717). Jerusalem: Israel Program for Scientific Translations [285].

Tomo, I., Kemper, C. M., and Lavery, T. J. (2010). Eighteen-year study of South Australian dolphins shows variation in lung nematodes by season, year, age class, and location. *Journal of Wildlife Diseases,* 46, 488 – 498 [286].

Torres, P., Oporto, J. A., Brieva, L. M., and Escare, L. (1992). Gastrointestinal helminths of the cetaceans *Phocoena spinipinnis* (Burmeister, 1865) and *Cephalorhynchus eutropia* (Gray, 1846) from the Southern coast of Chile. *Journal of Wildlife Diseases,* 28, 313 – 315 [287].

Toussant, P. R. A. (1977). Contribution à l'étude des facteurs de mortalité chez les cétacés des cotes de France. *Annales de la Societe des Sciences Naturelles de la Charente- Maritime,* 1 – 68 [288].

Travassos, L., Freitas, J. F. T., and Kohn, A. (1969). Trematódeos do Brasil. *Memórias do Instituto Oswaldo Cruz,* 67, 1 – 886 [289].

Treshchev, V. V. (1966). A new trematode *Orthosplanchus sudarikovi* nov. sp. (Campulidae) from the beluga. *Contribution to the Scientific Conference of the All-Union Society of Helminthologists, 3, 292–294.* [290].

Treshchev, V. V. (1968). The new campulid *Orthosplanchnus albamarinus* sp. n. (Trematoda, Campulidae) – parasite of the white whale. *Zoologichesky Zhurnal,* 47, 937 – 940 [291].

Treshchev, V. V. (1978). A study of trematodes from marine mammals in the European Arctic. *First All-Union Meeting of parasitocoenologists,* 3, 163 – 164 [292].

Treshchev, V. V. (1982). Specificity of helminths of marine mammals in the European section of the Arctic. In N. Dumka (Ed.), *Parazity i parazitozy cheloveka i zhivothnykh* (pp. 219 – 223). Kiev: Sbornik nauchnykh trudov [293].

Van Bree, P. J. H., Collet, A., Desportes, G., Hussenot, E., and Raga, J. A. (1986). Le dauphin the Fraser, *Lagenodelphis hosei* (Cetacea, Odontoceti), espéce nouvelle pour la faune d'Europe. *Mammalia,* 50, 57 – 86 [294].

Van Cleave, H. J. (1953). Acanthocephala of North American mammals. *Illinois Biological Monographs,* 23, 1 – 179 [295].

Van Thiel, P. H. (1966). The final hosts of the herringworm *Anisakis marina. Tropical and Geographical Medicine,* 18, 310 – 328 [296].

Van Waerebeek, K., Reyes, J. C., and Alfaro, J. (1993). Helminth parasites and phoronts of dusky dolphins *Lagenorhynchus obscurus* (Gray 1828) from Peru. *Aquatic Mammals,* 19, 159 – 169 [297].

Van Waerebeek, K., Reyes, J. C., Read, A. J., and McKinnon, J. S. (1990). Preliminary observations of bottlenose dolphins from the Pacific coast of South America. In S. Leatherwood, and R. R. Reeves (Eds.), *The Bottlenose Dolphin* (pp. 143 – 154). San Diego: Academic Press [298].

Viale, D. (1981). Lung pathology in stranded cetaceans on the Mediterranean coast. *Aquatic Mammals,* 8, 96 – 100 [299].

Walker, W. A. (2001). Geographical variation of the parasite, *Phyllobothrium delphini* (Cestoda), in Dall's porpoise, *Phocoenoides dalli,* in the Northern North Pacific, Bering Sea, and Sea of Okhotsk. *Marine Mammal Science,* 17, 264 – 275 [300].

Wazura, K. W., Strong, J. T., Glenn, C. L., and Bush, A. O. (1986). Helminths of the beluga whale (*Delphinapterus leucas*) from the Mackenzie River Delta, Northwest Territories. *Journal of Wildlife Diseases,* 22, 440 – 442 [301].

Webster, W. A., Neufeld, J. L., and MacNeill, A. C. (1973). *Halocercus monoceris* sp. n. (Nematoda: Metastrongyloidea) from the Narwhal, *Monodon monoceros. Proceedings of the Helminthological Society of Washington*, 40, 255 – 258 [302].

Williams, H. H. (1968). The taxonomy, ecology, and host specificity of some Phyllobothriidae (Cestoda: Tetraphyllidea), a critical revision of *Phyllobothrium* Beneden, 1849 and comments on some allied genera.

Philosophical Transactions of the Royal Society B: Biological Sciences, 253, 231 – 307 [303].

Woodard, J. C., Zam, S. G., Caldwell, D. K., and Caldwell, M. C. (1969). Some parasitic diseases of dolphins. *Pathologia Veterinaria,* 6, 257 – 272 [304].

WoRMS Editorial Board. (2015). World Register of Marine Species. Accessed through: www.marinespecies.org at VLIZ on 2015 – 08 – 01.

Wulker, G. (1930). Über Nematoden aus Nordseetieren I. *Zoologischer Anzeiger,* 87, 293 – 303 [305].

Yablokov, A. V., Bel'kovich, V. M., and Borisov, V. I. (1972). *Whales and dolphins, part I.* Moscow: Kity I Del'finy. Nauk SSSR [306].

Yamaguti, S. (1939). Studies on the helminth fauna of Japan. Part 29. Acanthocephala, II. *Japan Journal of Zoology,* 8, 317 – 351 [307].

Yamaguti, S. (1942). Studies on the helminth fauna of Japan. Part 40. Three new species of trematodes from the bile ducts of marine mammals. *Transactions of the Biogeographical Society of Japan,* 3, 390 – 407 [308].

Yamaguti, S. (1951a). Studies on the helminth fauna of Japan. Part 45. Trematodes of marine mammals. *Acta Medica Okayama,* 7, 283 – 294 [309].

Yamaguti, S. (1951b). Studies on the helminth fauna of Japan Part 46. Nematodes of marine mammals. *Acta Medica Okayama,* 7, 295 – 306. [310].

Yamaguti, S. (1951c). Studies on the helminth fauna of Japan Part 47. Cestodes of marine mammals and birds. *Acta Medica Okayama,* 7, 307 – 316 [311].

Yamaguti, S. (1959). *Systema Helminthium, Vol. II. The Cestoda of Vertebrates*. New York: Interscience. [312].

Yamaguti, S. (1963). *Systema Helminthum Vol. V Acanthocephala*. New York: Interscience [313].

Yamaguti, S. (1971). *Synopsis of digenetic trematodes of vertebrates, Vol. I.* Tokyo: Keigaku Publishing Company [314].

Yazaki, S., Kamo, H., Maejima, J., and Yamane, Y. (1979). On a species of the genus *Plicobothrium* expelled from a shortfinned balckfish, *Globicephala macrorhynchus* (Gray, 1846) in Toba Aquarium, Japan. *Yonago Acta Medica,* 23, 21 – 25 [315].

Yurachno, M. V. (1967). *Orthosplanchnus pygmaeus* sp. n. (Trematoda, Campulidae) a parasite of the whale. *Vestnik Zoologii,* 3, 79 – 82 [316].

Zam S. G., Caldwell D. K., and Caldwell M. C. (1970). Some internal parasites from freshwater cetaceans from the upper Amazon River. *Investigations on Cetacea,* 2, 250 – 251 [317].

Zam S. G., Caldwell D. K., and Caldwell M. C. (1971). Some endoparasites from small odontocete cetaceans collected in Florida and Georgia. *Cetology,* 2, 1 – 11 [318].

Zerbini, A. N., and Santos, M. C. O. (1997). First record of the pygmy killer whale *Feresa attenuata* (Gray, 1874) for the Brazilian coast. *Aquatic Mammals,* 23, 105 – 109 [319].

Zylber, M. I., Failla, G., and Le Bas, A. (2002). *Stenurus globicephalae* Baylis et Daubney, 1925 (Nematoda: Pseudaliidae) from a false killer whale, *Pseudorca crassidens* (Cetacea: Delphinidae), stranded on the coast of Uruguay. *Memórias do Instituto Oswaldo Cruz, Rio de Janeiro,* 97, 221 – 225 [320].

In: Advances in Marine Biology. Volume 1 ISBN: 978-1-53610-131-7
Editors: Adam Kovács and Patrik Nagy © 2016 Nova Science Publishers, Inc.

Chapter 4

EFFECT OF CLAYS ON THE THE REMOVAL OF OKADAIC ACID FROM SEAWATER

S. Lucía Blanco[1], Laura P. Rodríguez[1], José Luis Legido[2], Carmen P. Gómez[3], Juan M. Vieites[1] and Ana G. Cabado[1,*]

[1]Food Safety Division- IDi.ANFACO-CECOPESCA, Carretera Colexio Universitario (Pontevedra), Spain
[2]Departamento de Física Aplicada. Facultad de Ciencias del Mar. Universidad de Vigo. Vigo, Spain
[3]Peloides Termales S.L., Vigo, Spain

ABSTRACT

The use of clays for mitigating the effects of toxic episodes causing continuous shellfish harvesting closures owing to lipophilic toxins remains unexplored. This strategy is promising for clearance of microalgae through flocculation, but few studies evaluate toxins elimination. This study presents preliminary data on the withdrawal of okadaic acid, testing different clays in seawater, in laboratory conditions. Clays were assayed based on the potential effectiveness in terms of specific surface, particle size and colloidal properties, as well as their local abundance: three kinds of bentonites (one containing sodium

* Author for correspondence: E-mail: agcabado@anfaco.es. Tel.: +34986469303, Fax: +34986469269

bentonite particles <70 μm, other, 83% < 2 μm, and bentonite mainly composed of muscovite, particles >45 μm) and kaolinites of different origins and grain size (69% < 2 μm and 57% < 2 μm). Bentonite (83% < 2 μm) had the best toxin removal ability in our study (68%), while kaolinites with longer decantation times (69% < 2 μm) were the least effective material. The particles with the highest size (bentonite >45 μm) showed a lower efficiency, probably related to the fast decantation process. Altogether, clay applications offer an opportunity for toxin removal that should be explored in depth.

1. Introduction

Harmful algal blooms (HABs) pose a serious concern in many coastal zones all over the world, especially those where aquaculture activities are relevant. In the northwest of Spain, Galicia, where the Rías conform a particular habitat supporting aquaculture and fishing activities, toxic episodes are associated to atmospheric and oceanographic interactions, specifically upwelling and downwelling, and transitions between water column stability and vertical mixture. When phytoplanktonic toxin producer species are transported and aggregated influenced by oceanographic conditions, and their trophic needs are covered by the abundance of prey species, massive proliferations of toxic phytoplankton occur [1]. One of the species of concern in Galicia is *Dinophysis* spp., a dinoflagellate responsible for lipophilic toxins production, formerly known as "diarrhetic" toxins, because they trigger the Diarrhetic Shellfish Poisoning (DSP) when contaminated mollusks are consumed. Okadaic acid and dinophysistoxins are the main analogues involved in toxic episodes in Galicia, causing closures in harvesting areas and producing an economical detriment of the bivalve aquaculture sector. Mussels growing in rafts in floating settlements are the most affected species, when they ingest the toxins by filter feeding. Although they do not suffer adverse effects, several weeks are necessary for depuration. At present, this natural elimination of toxins in the marine environment is the only allowed process in shellfish aquaculture.

Closure periods with prohibition of mollusk extraction, have grown in frequency and duration during the last years [2]. Although the spring and the autumn are the main periods of toxic episodes, closures have been extended during 4 consecutive months, or even more, in the worst cases. Estimation of costs associated to the toxic episodes in Galicia ascended to 40 million € in 2010 [3].

Clays have been used in control of harmful algal blooms (HABs) in Asia [4, 5], laboratory and field studies have been performed in several countries over the past 30 years. Some assays reproducing environmental conditions have been performed with the aim of using these kind of materials for mitigation of toxic blooms [6, 7], and clay dispersal was tested in situ in Japan, Korea and Australia [8].

Clearance of microalgae is possible through flocculation processes by use of mineral particles. This mechanism poses a promising strategy for control of the harmful algal blooms and for mitigating the impact on the local resources [9]. Clay dispersal and flocculation has been used in Korea for mitigation of *Cochlodinium polykrikoides* blooms, a dinoflagellate causing fish and shellfish mortality, as a strategy included in the management scheme [8]. Experiments have been done with *Karenia brevis,* a toxic dinoflagellate producing brevetoxins, responsible for mortalities of fish, invertebrates and marine mammals in coastal waters of Florida and Gulf of Mexico [10]; *Prymnesium parvum,* a potent fish killing organism that produces prymnesins, commonly found in low salinity or mesohaline waters [5, 11], and more recently, with paralytic shellfish poisoning causative organisms in the Philippines [12], among others. Nevertheless, few studies have been performed to evaluate the toxins removal when applying these strategies.

Some reports have shown that the cyanobacterial hepatotoxin microcystin-LR can be adsorbed by clay particles, enabling the removal of 81% of this toxin from the water [13], and also the removal of prymnesins [5], and 70% of extra-cellular brevetoxins, from *Karenia brevis,* in seawater [14]. Natural clays and its composites are known to trap pollutants by taking up cations and anions through ion exchange and/or adsorption, removing different organic and inorganic contaminants from water. Moreover, it has been proved that in many cases clays are better or comparable with commercial filter materials for drinking water, with the advantage of being a low-cost green, nontoxic adsorbent [15]. Some recent studies focus on modified clays or even modified sand [16, 17] and their effects on algal cells and toxins, with the aim of maximizing effectiveness, and to minimize costs and environmental impact. Some examples are assays for removal of microcystins from drinking water using nano-sized montmorillonite K10 [18, 19], nanosilicate platelet derived from natural clay mineral [20], kaolin modified with polyaluminium chloride for removing *Alexandrium tamarense* and paralytic shellfish poisoning toxins [21]. Other recent approach is the development of clay-mimicking materials, such as aminoclays, with algicidal activity [22].

To our knowledge, no studies have been published regarding the effect of natural clays on the okadaic acid group of lipophilic toxins. During proliferations of *Dinophysis*, an important proportion of their toxins are released into the seawater [23]. Mussels ingest toxic phytoplankton and accumulate lipophilic toxins mainly in the digestive gland [24]. Recently, it was described that dissolved azaspiracids are absorbed and metabolized by blue mussels, showing that toxins uptake from the dissolved phase through the gills is a possible route of accumulation [25], although no results on accumulation by the gills have been reported using only dissolved lipophilic toxins. It was suggested that mussels can accumulate diarrhetic toxins dissolved in the seawater, from an experimental study [26], but these results have not been confirmed from field studies. Dissolved toxins may pose a further risk to shellfish farms if mussels are able to take up the toxins in the dissolved form in the same manner as ingesting toxic algae.

This study proposes testing different clays in seawater naturally contaminated with okadaic acid from a toxic episode in the Galician coast and obtain a preliminary evaluation of the possible removal of this toxin from seawater in laboratory conditions, with the aim of providing evidence that further research would be of interest.

Different clays were selected based on the potential effectiveness in terms of specific surface, particle size and colloidal properties, as well as their local abundance. Bentonites have been shown in former studies to be effective against a broad range of algal species, due to their ability to expand in water [5]. Then, three varieties of bentonite were assayed: "Bentonite P" was sodium bentonite, and contained particles <70 μm, "Bentonite C" (83% < 2 μm) and "Bentonite > 45 μm", the mineral composed mainly by muscovite, with particle size > 45 μm. Other compounds tested in this preliminary study were local clays ("Grove", 69% < 2 μm, and "Lendo", 57% < 2 μm) mainly composed by kaolinite.

2. Materials and Methods

2.1. Selection of Clays

Clays were provided by two Galician companies: "Caolines de Vimianzo, S.A." (CAVISA) a company that commercializes kaolins, micas and sands; and "Peloides Termales", that designs products for balneotherapy and thalassotherapy. Clays were selected for the specific surface and the particle size.

The clay supplied by Peloides Termales, "Bentonite P" was sodium bentonite, and contained particles <70 μm. The studied sample showed a high percentage of phyllosilicates in the total fraction (98%). The clay fraction included the following minerals in decreasing amounts: smectite (56%), sepiolite (29%) and illite (15%) [27, 28]. Sodium bentonite expands when wet, absorbing as much as several times its dry mass in water. It was selected because of its excellent colloidal properties.

The clays supplied by CAVISA were "Bentonite C" (83% < 2 μm), "Bentonite > 45 μm" (mainly muscovite) and kaolinites locally available from the regions Lendo and Grove, near the Galician coast: "Lendo Clay" (57% < 2 μm), "Grove Clay" (69% < 2 μm). Figure 1 shows size distribution (granulometry) of the clays.

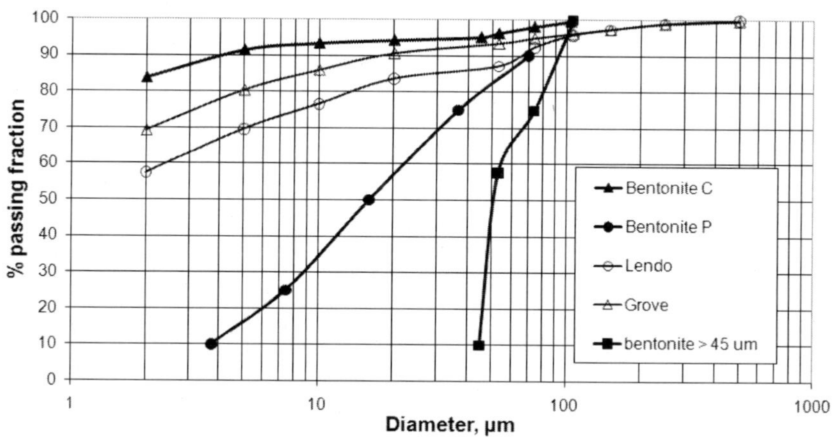

Figure 1. Granulometry of the clays selected in the study. The "passing fraction" is the total percentage of clay that passes through the sieve of a given size (indicated by parameter "diameter").

2.2. Seawater Sampling

Seawater from the Galician coast was sampled during a toxic episode in Ría de Aldán, Galician coast, northwest of Spain (Figure 3) at a depth of about 1 meter, and kept frozen in plastic containers. The species involved in this toxic episode was *Dinophysis acuminata*, registered at densities of 2000-6000 cells/liter in the 10-day period around the sampling date. The harvesting area was closed, with okadaic acid levels higher than the legal limit (>160 μg / kg) in the mollusks. Okadaic acid content was measured using the enzyme-linked immunosorbent assay (ELISA) from Abraxis; the obtained value was 0.37 ± 0.05 ng / mL (n = 11) (mean value ± SEM). These results were obtained before and after freezing, and analyses were done at different times, during a period of several months. Figure 2 shows the distribution of the obtained values of okadaic acid in seawater throughout the study.

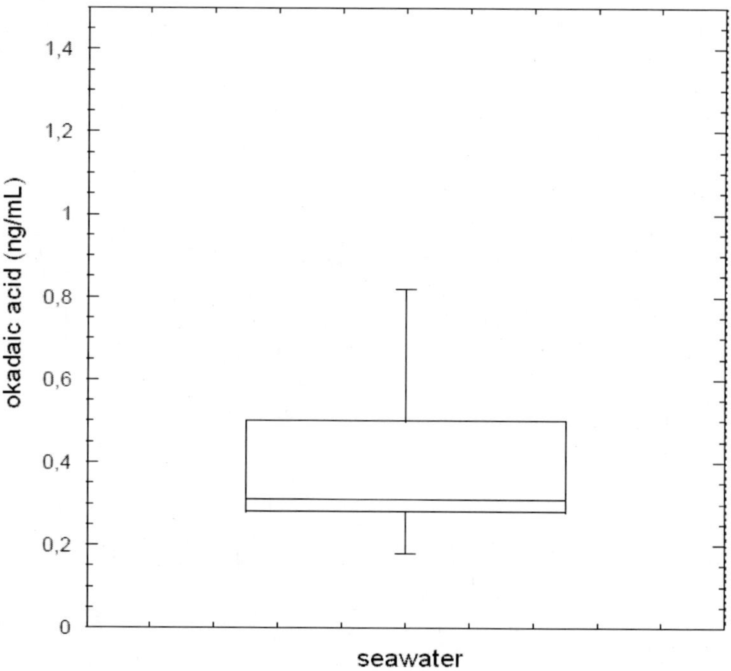

Figure 2. Distribution of values of okadaic acid in seawater during the study.

2.3. Clay Suspensions Preparation

The clays concentrations tested in the present study were in the range from 2.5 to 15.0 g/L. Seawater and clays suspensions were prepared as follows: 50 mL of water were dispensed in test tubes and the corresponding clay was added to achieve 5 different concentrations (0.25, 0.50, 0.75, 1.00 and 1.50% clay) in the seawater. The mixtures were allowed to settle and separate by gravity. Decantation took place in the tubes, containing each clay/concentration. Seawater subsamples were taken at different sampling times (12, 24, 48 and 72 hours) using a syringe and taking care for not disturbing the sediment. Water samples obtained were analyzed for okadaic acid.

2.4 Okadaic Acid Analysis

Enzyme-Linked Immunosorbent Assay (ELISA) from Abraxis (Warminster, Pennsylvania, USA) was used for the determination of okadaic acid. The test was performed following specifications of the manufacturer. The final color was evaluated using a spectrometer Spectramax M5 microplate reader (Molecular Devices Inc., Sunnyvale, CA, USA) at 450 nm, and the concentrations of the samples were determined by interpolation using the standard curve constructed with each run, covering the range of concentrations from 0.1 to 5.0 ng / mL.

3. Results

First, we tested the effect of the different clays, by preparing suspensions in seawater collected during a toxic episode in Ría de Aldán, Galician coast (Figure 3).

Several clay concentrations were prepared, in test tubes (0.25, 0.50, 0.75, 1.00 and 1.50%) and allowed to stand, until decantation (allowing the mixture to settle and separate by gravity). Seawater was extracted at different sampling times: 12, 24, 48 and 72 hours.

Results obtained after experiments with clays are shown in Figures 4 - 8. Mean value of duplicates obtained in the ELISA test are represented; the coefficients of variation for replicates are <5% in most cases, and <15% in all of them, within the method specifications.

A decrease in the okadaic acid content of the seawater after decantation of clay was registered (Figures 4-8). Figure 4 shows variations in okadaic acid content from seawater after addition of clay "Bentonite P" (<70 μm), results of replicate analyses are shown (mean values ± Standard Error of the Mean (SEM)).

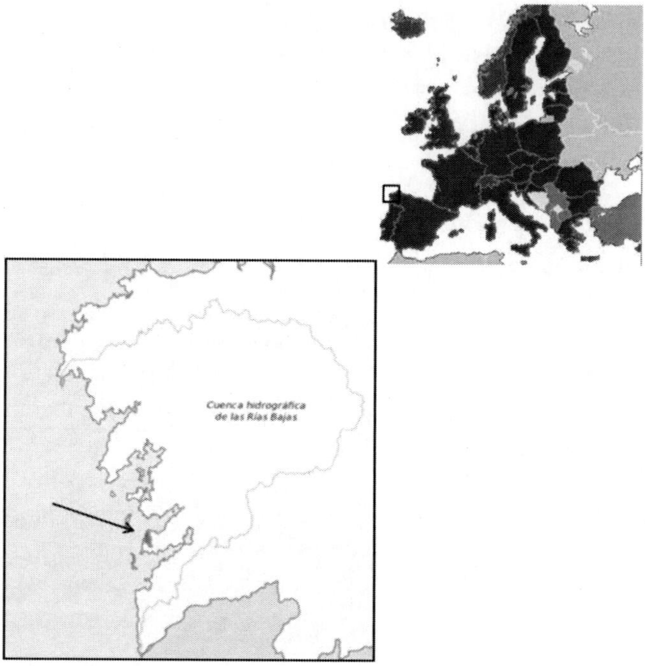

Figure 3. Sampling site for seawater during a toxic episode in Ría de Aldán (Galician Coast, Spain).

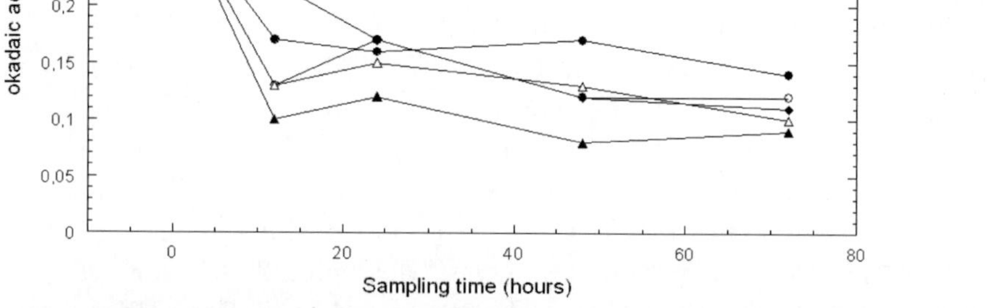

Figure 4. Variations in the okadaic acid content from seawater after addition of clay "Bentonite P" (<70 μm) at different concentrations (0.25 – 1.50%) and sampling times (12 – 72 hours).

The decrease of okadaic acid in the seawater after addition of clay was evident after the first 12 hours with Bentonite P (<70 μm) (Figure 4) and Bentonite C (83% < 2 μm) at all the concentrations tested, obtaining a mean decrease of 68% (8% CV) (Figure 5).

With kaolinitic clays from Lendo and Grove, results were also promising, although more variations were observed. At 24 hours, the reduction of okadaic acid was maximum when Lendo clay (57% < 2 μm) (Figure 6) and Grove clay (69% < 2 μm) (Figure 7) were used at all the concentrations tested, 66 and 67% in average, respectively (8 and 13% CV). The increase of okadaic acid observed at 48 h (from 0.1 or < 0.1 ng / mL at 24 h to values of 0.26 - 0.28 ng / mL at 48 h) at the lower concentrations (<0.75%) of Grove clay may be related to the lower particle size of Grove clay (69% < 2 μm), compared to Lendo clay (57% < 2 μm), involving a possible resuspension or not enough "attachment" of the toxin to the deposited clay.

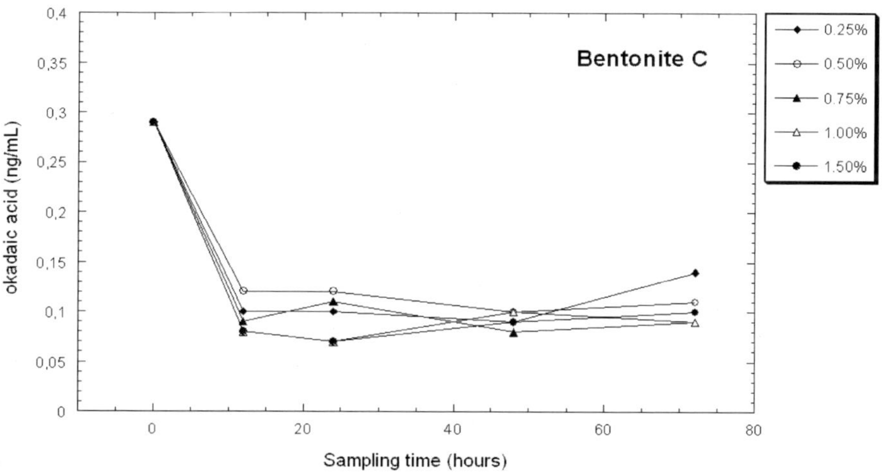

Figure 5. Variations in the okadaic acid content from seawater after addition of clay "Bentonite C" (83% < 2 μm) at different concentrations (0.25 – 1.50%) and sampling times (12 – 72 hours).

This effect was not observed at higher concentrations of this clay, possibly owing to a more stable settlement of this higher amount of particles and "attached" molecules on the bottom. Higher values than the initial okadaic acid concentration obtained at 12 hours in both clays seemed to be experimental artifacts, probably related with longer decantation times, and hence longer time in suspension in the seawater, possibly causing interference in the ELISA measurement.

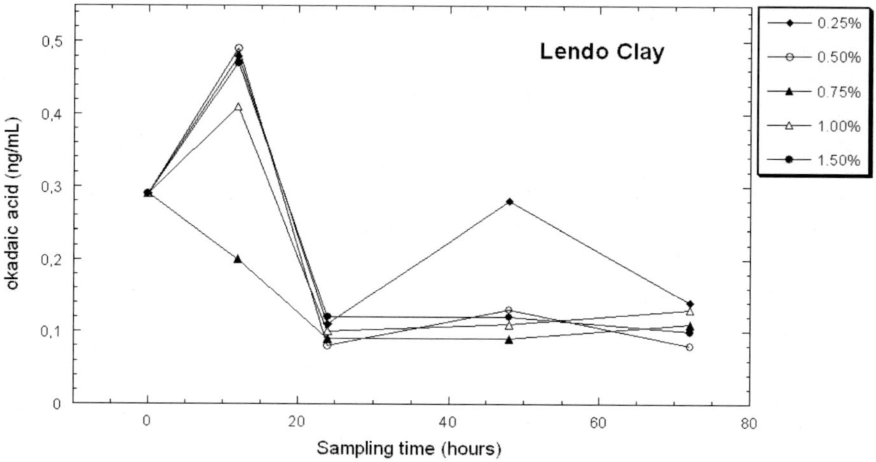

Figure 6. Variations in the okadaic acid content from seawater after addition of kaolinitic "Lendo clay"(57% < 2 μm) at different concentrations (0.25 – 1.50%) and sampling times (12 – 72 hours).

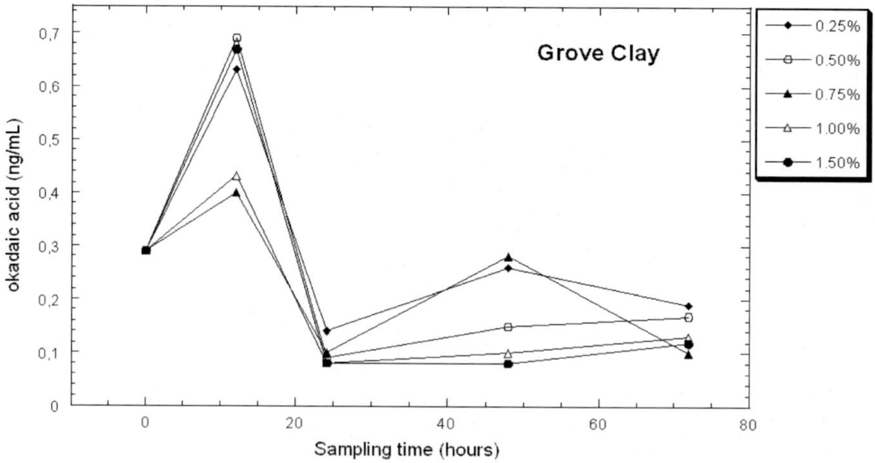

Figure 7. Variations in the okadaic acid content from seawater after addition of kaolinitic "Grove clay"(69% < 2 μm) at different concentrations (0.25 – 1.50%) and sampling times (12 – 72 hours).

Both bentonites (83% < 2 μm and <70 μm, respectively) and kaolinitic Lendo clay (57% < 2 μm) showed a high degree of decantation 2 hours after the suspensions preparation, in the laboratory conditions. By contrast, kaolinitic Grove clay (69% < 2 μm) needed about 22 h for decantation and clarification of the solutions, staying in suspension a longer time than the remaining clays assayed in this study.

The highest particle size material, bentonite >45 μm, also caused a reduction in seawater OA content (Figure 8), but at longer exposure times of the experiment, 48-72 hours, and to a lesser extent, about 43-46%, respectively. Longer time was needed for toxin removal from the seawater, at concentrations over 0.25%. This material was totally decanted after the first 10 minutes of suspension preparation.

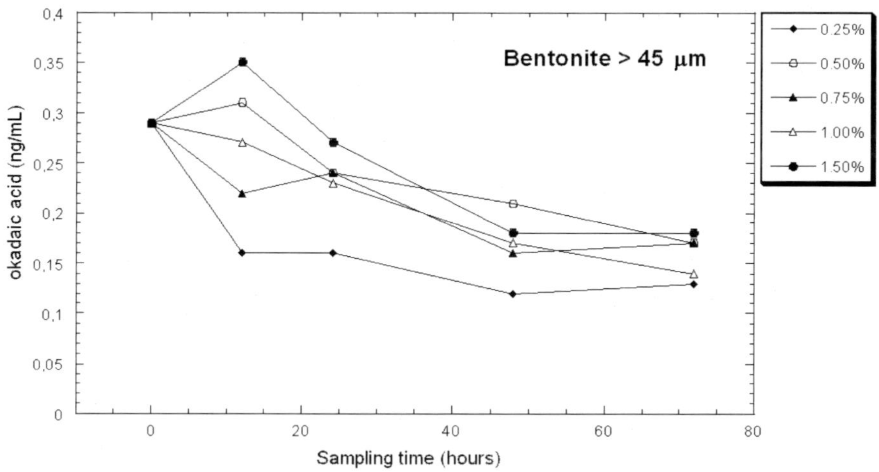

Figure 8. Variations in the okadaic acid content from seawater after addition of clay "Bentonite >45 μm" at different concentrations (0.25 – 1.50%) and sampling times (12 – 72 hours).

In summary, the most promising results of toxin removal from seawater obtained with clays, were obtained at 12 hours after the addition of bentonites C (83% < 2 μm) and P (<70 μm), 24 hours for kaolinitic clays 57% and 69% < 2 μm (Lendo and Grove clays), and 72 hours with bentonite >45 μm. Bentonite C (83% < 2 μm) seems to offer the highest decrease of okadaic acid in seawater.

4. DISCUSSION

Recovery of hydrophobic toxins from seawater by adsorption of toxins onto hydrophobic substances is usually carried out for analyses. It was shown that the finest particles of marine sediment consisting mainly of montmorillonite adsorbed oleic acid (references cited by [13]). Clay properties of fine-grained particles, high surface area and charge suggest that this material could be effective for adsorbing hydrophobic molecules from water, as proposed by others for microcystins and extra cellular brevetoxin molecules [13, 14]. Okadaic acid and dinophysistoxins are fat soluble, have high molecular weights, and belong to a class of compounds called polycyclic ethers. These compounds are similar in chemical structure to other marine toxins such as yessotoxins, brevetoxins, ciguatoxins and palytoxins [29]. Okadaic acid consists structurally in a relatively rigid hydrophobic tail structure connected via a flexible arm to a very rigid spiroketal structure with an acidic group, capable of coordination to the metals and/or the aquo ligands at the active site of the enzyme protein phosphatase-1 [30]. Toxins are associated in most cases with the presence of their producers, inside algal cells, but in some cases, toxins may be present in the water column associated with suspended matter and bottom sediments or water bodies [31]. It is known that microcystins, and especially anatoxin-a, adsorb to fine clay material, and can accumulate in the sediment; a recent study found that okadaic acid was detected at remarkable levels in suspended solids during *Dinophysis* bloom in Alfacs Bay, one of Ebro´s associated marine bays [31]. The present study intends to explore the use of clays for mitigating the effects of the toxic episodes in Galicia, especially those originating from *Dinophysis* sp., dinoflagellates responsible for the production of the formerly known as "Diarrhetic Shellfish Poisoning" toxins, or DSP toxins. The incidence of these lipophilic toxins in the mussel and other bivalve mollusks aquaculture sector is increasing in the last years, globally, possibly influenced by eutrophication, maritime transport and climatic change [2]. Bentonites have been effective in previous studies for a broad range of algal species, due to their ability to expand in water, and to develop strong surface change [5]. As observed by others, bentonite (83% < 2 μm) had the best removal ability in our study (68%). Theoretically, larger particle size would increase the rate of particle collision, while a stronger surface charge would lead to a higher proportion of particle adhesion on contact, and both processes would lead to higher flocculation rates [5]. On the other hand, kaolinite (43% < 2 μm) was the least effective material, only showing a remarkable decrease of okadaic acid at 24 h

(58%), but increasing again afterwards. It was described that kaolinites did not swell when wetted, and had a much lower charge density than bentonites [5], so, agreeing with previous studies, they were less effective in toxin removal. Particle concentration is another important factor, as described by others, in flocculation rates, along with particle size and charge [32]. This effect might explain the behavior of the kaolinitic clays locally available from Lendo and Grove, with a high proportion of particles <2 μm, and with better results in okadaic acid removal from seawater at higher concentrations of both clays, especially in Grove clay, with a higher proportion of small particles (69% < 2 μm) than Lendo clay (57% < 2 μm)). It has been suggested that when aggregates are >1 μm, differential sedimentation, i.e., collisions resulting from the interception of smaller particles by larger, more-rapidly sinking particles, is the dominant transport mechanism (reported by [33]). The effectiveness of differential sedimentation may also diminish in time as the aggregates attain a certain size and displace increasing amounts of water during their descent, which can push away smaller particles instead of colliding and attaching with them, phenomenon known as hydrodynamic retardation [33]. This effect would explain the behavior of the clay mainly composed of muscovite, bentonite >45 μm, at concentrations higher than 0.25%, with lower efficiency of toxin removal from the seawater. As expected, practically complete decantation of these particles occurred within the first 10 minutes, after the suspension preparation. Use of clays in the field pose environmental concerns, since the toxic cells and/or toxins sedimentation could lead to toxin accumulation by benthic organisms [34], and also impacts of clay on sessile organisms and other co-occurring species should be evaluated. Nevertheless, the number of studies, although limited, shows only low or no impact on co-occurring species, and a negative effect on bivalve growth [35]. On the other hand, clay treatment would not be recommended if flocculation required several hours to take effect, as no water column would be static for such a length of time in nature [5]. Then, the materials composed by kaolinite, would not be the most adequate for achieving okadaic acid toxins removal from the water column. Further studies assaying shorter periods of time, lower clay concentration and tests with the causative organisms, *Dinophysis* sp. and *Prorocentrum* sp. would be desirable.

In summary, use of clays as a mitigation strategy for microalgal toxins contamination offered promising results. However, further research on the toxin and/or causative agent removal efficiency and the environmental implications of this strategy should be evaluated.

CONCLUSION

Bentonite (83% < 2 μm) had the best okadaic acid removal ability from seawater in our study (68%). By contrast, kaolinite (43% < 2 μm) was the least effective material, only showing a remarkable decrease of okadaic acid at 24 h (58%), but increasing again afterwards. The other kaolinite clays, from Lendo (57% < 2 μm) and from Grove (69% < 2 μm), also showed promising results, although longer decantation times in Grove clay and kaolinite limit their possible usefulness.

The particles with the highest size (bentonite >45 μm) showed a lower efficiency, probably related to the fast decantation process. Altogether, clay applications offer an opportunity for toxin removal that should be explored in depth.

ACKNOWLEDGMENTS

The Authors are grateful to Jesús Pais, from Caolines de Vimianzo S.A.U. (CAVISA), Vimianzo, A Coruña, Spain, for providing the clay materials for the study, and to María Perfecta Salgado González, for collaborating in the experimental setup. The research leading to these results has received funding from the following FEDER cofounded-grants: Centre for Industrial Technological Development (CDTI) and Technological Funds, supported by Ministerio de Economía y Competitividad, AGL2012-40185-CO2-01 and through the Galician Agency of Innovation (GAIN), Spain ITC-20133020 SinTox, supported by the program FEDER-INNTERCONECTA by GAIN and CDTI, and cofunded by the Technology Fund.

REFERENCES

[1] Velo-Suárez, L.; González-Gil, S.; Pazos, Y.; Reguera, B. The growth season of Dinophysis acuminata in an upwelling system embayment: A conceptual model based on in situ measurements. *Deep Sea Research Part II: Topical Studies in Oceanography* 2014, *101*, 141-151.

[2] James, K. J.; Carey, B.; O'Halloran, J.; van Pelt, F. N. A. M.; SkrabáKová, Z. Shellfish toxicity: human health implications of marine algal toxins. *Epidemiology and Infection* 2010, *138*, 927-940.

[3] PURGADEMAR. *Sistema de Observación y Alerta de Proliferación de Microalgas Nocivas. Available online: http://www.purgademar.com/ (accesed on 14 May 2015).*

[4] Kim, H. G. Cochlodinium polykrikoides blooms in Korean coastal waters and their mitigation. In: B. Reguera, J. Blanco, M.L. Fernández, and T. Wyatt (eds.), Harmful algae. Xunta de Galicia and IOC of UNESCO, Santiago de Compostela. 1998, 227-228.

[5] Sengco, M. R.; Hagstrom, J. A.; Graneli, E.; Anderson, D. M. Removal of *Prymnesium parvum* (Haptophyceae) and its toxins using clay minerals. *Harmful Algae* 2005, *4*, 261-274.

[6] Beaulieu, S. E.; Sengco, M. R.; Anderson, D. M. Using clay to control harmful algal blooms: deposition and resuspension of clay/algal flocs. *Harmful Algae* 2005, *4*, 123-138.

[7] Orizar, I. S.; Rivera, P. P. L.; San Diego-McGlone, M. L.; Azanza, R. V. Harmful Algal Bloom (HAB) mitigation using ball clay: Effect on non-target organisms. *Journal of Environmental Science and Management* 2013, 36-43.

[8] Park, T. G.; Lim, W. A.; Park, Y. T.; Lee, C. K.; Jeong, H. J. Economic impact, management and mitigation of red tides in Korea. *Harmful Algae* 2013, *30*, S131-S143.

[9] Sengco, M. R.; Anderson, D. M. Controlling harmful algal blooms through clay flocculation. *J. Eukaryot. Microbiol.* 2004, *51*, 169-172.

[10] Lewis, M. A.; Dantin, D. D.; Walker, C. C.; Kurtz, J. C.; Greene, R. M. Toxicity of clay flocculation of the toxic dinoflagellate, Karenia brevis, to estuarine invertebrates and fish. *Harmful Algae* 2003, *2*, 235-246.

[11] Hagstrom, J. A.; Graneli, E. Removal of Prymnesium parvum (Haptophyceae) cells under different nutrient conditions by clay. *Harmful Algae* 2005, *4*, 249-260.

[12] Padilla, L. V.; San Diego-McGlone, M. L.; Azanza, R. V. Exploring the potential of clay in mitigating *Pyrodinium bahamense* var. *compressum* and other harmful algal species in the Philippines. *Journal of Applied Phycology* 2010, *22*, 761-768.

[13] Morris, R. J.; Williams, D. E.; Luu, H. A.; Holmes, C. F. B.; Andersen, R. J.; Calvert, S. E. The adsorption of microcystin-LR by natural clay particles. *Toxicon* 2000, *38*, 303-308.

[14] Pierce, R. H.; Henry, M. S.; Higham, C. J.; Blum, P.; Sengco, M. R.; Anderson, D. M. Removal of harmful algal cells (*Karenia brevis*) and toxins from seawater culture by clay flocculation. *Harmful Algae* 2004, *3*, 141-148.

[15] Srinivasan, R. Advances in Application of Natural Clay and Its Composites in Removal of Biological, Organic, and Inorganic Contaminants from Drinking Water. *Advances in Materials Science and Engineering* 2011.

[16] Li, L.; Pan, G. A Universal Method for Flocculating Harmful Algal Blooms in Marine and Fresh Waters Using Modified Sand. *Environmental Science and Technology* 2013, *47*, 4555-4562.

[17] Pan, G.; Chen, J.; Anderson, D. M. Modified local sands for the mitigation of harmful algal blooms. *Harmful Algae* 2011, *10*, 381-387.

[18] Couto, R. S. P.; Guarino, A. W. S.; Branco, C. W. C.; Palermo, E. F. A.; Azero, E. G. Application of Clay Minerals and Polymeric Resins to Remove Dissolved Microcystin-LR from Water. *International Journal of Environmental Research* 2013, *7*, 435-442.

[19] Wang, Z. Y.; Wang, C.; Wang, P. F.; Qian, J.; Hou, J.; Ao, Y. H. Process Optimization for Microcystin-LR Adsorption onto Nano-sized Montmorillonite K10: Application of Response Surface Methodology. *Water Air and Soil Pollution* 2014, *225*.

[20] Chang, S. C.; Li, C. H.; Lin, J. J.; Li, Y. H.; Lee, M. R. Effective removal of Microcystis aeruginosa and microcystin-LR using nanosilicate platelets. *Chemosphere* 2014, *99*, 49-55.

[21] Lu, G. Y.; Song, X. X.; Yu, Z. M.; Cao, X. H.; Yuan, Y. Q. Environmental effects of modified clay flocculation on Alexandrium tamarense and paralytic shellfish poisoning toxins (PSTs). *Chemosphere* 2015, *127*, 188-194.

[22] Lee, Y. C.; Jin, E.; Jung, S. W.; Kim, Y. M.; Chang, K. S.; Yang, J. W.; Kim, S. W.; Kim, Y. O.; Shin, H. J. Utilizing the algicidal activity of aminoclay as a practical treatment for toxic red tides. *Scientific Reports* 2013, *3: 1292 DOI: 10.1038/srep01292*.

[23] MacKenzie, L.; Beuzenberg, V.; Holland, P.; McNabb, P.; Selwood, A. Solid phase adsorption toxin tracking (SPATT): a new monitoring tool that simulates the biotoxin contamination of filter feeding bivalves. *Toxicon* 2004, *44*, 901-918.

[24] Blanco, J., Mariño, C., Martín, H., Acosta, C. P. Anatomical distribution of Diarrhetic Shellfish Poisoning (DSP) toxins in the mussel *Mytilus galloprovincialis*. *Toxicon* 2007, *50(8)*, 1011-1018.

[25] Jauffrais, T.; Kilcoyne, J.; Herrenknecht, C.; Truquet, P.; Sechet, V.; Miles, C. O.; Hess, P. Dissolved azaspiracids are absorbed and metabolized by blue mussels (*Mytilus edulis*). *Toxicon* 2013, *65*, 81-89.

[26] Wallace, Glenn Manfred. 2011. "Diarrhetic Shellfish Toxins in Tasmanian coastal waters: causative dinoflagellate organisms, dissolved toxins and shellfish depuration." PhD diss., University of Tasmania.
[27] Casás, L. M.; Legido, J. L.; Pozo, M.; Mourelle, L.; Plantier, F.; Bessières, D. Specific heat of mixtures of bentonitic clay with sea water or distilled water for their use in thermotherapy. *Thermochimica Acta* 2011, *524*, 68-73.
[28] Casás, L. M.; Pozo, M.; Gómez, C. P.; Pozo, E.; Bessières, L. D.; Plantier, F.; Legido, J. L. Thermal behavior of mixtures of bentonitic clay and saline solutions. *Applied Clay Science* 2013, *72*, 18-25.
[29] Dominguez, H. J.; Crespin, G. D.; Santiago-Benitez, A. J.; Gavin, J. A.; Norte, M.; Fernandez, J. J.; Daranas, A. H. Stereochemistry of Complex Marine Natural Products by Quantum Mechanical Calculations of NMR Chemical Shifts: Solvent and Conformational Effects on Okadaic Acid. *Marine Drugs* 2014, *12*, 176-192.
[30] Lindvall, M. K.; Pihko, P. M.; Koskinen, A. M. P. The binding mode of calyculin A to protein phosphatase-1 - A novel spiroketal vector model. *Journal of Biological Chemistry* 1997, *272*, 23312-23316.
[31] Rivetti, C.; Gomez-Canela, C.; Lacorte, S.; Barata, C. Liquid chromatography coupled with tandem mass spectrometry to characterise trace levels of cyanobacteria and dinoflagellate toxins in suspended solids and sediments. *Analytical and Bioanalytical Chemistry* 2015, *407*, 1451-1462.
[32] Yu, Z. M.; Zou, J.Z.; Ma, X.N. Application of clays to removal of red tide organisms I. coagulation of red tide organisms with clays. *Chinese Journal of Oceanology and Limnology* 1994, *12*, 193-200.
[33] Yu, Z. M.; Sengco, M. R.; Anderson, D. M. Flocculation and removal of the brown tide organism, Aureococcus anophagefferens (Chrysophyceae), using clays. *Journal of Applied Phycology* 2004, *16*, 101-110.
[34] Haubois, A. G.; Bricelj, V. M.; Naar, J. Transfer of brevetoxins to a tellinid bivalve by suspension- and deposit-feeding and its implications for clay mitigation of *Karenia brevis* blooms. *Marine Biology* 2007, *151*, 2003-2012.
[35] Hagstrom, J. A.; Sengco, M. R.; Villareal, T. A. Potential Methods for Managing *Prymnesium Parvum* Blooms and Toxicity, with Emphasis on Clay and Barley Straw: a Review. *Journal of the American Water Resources Association* 2010, *46*, 187-198.

In: Advances in Marine Biology. Volume 1 ISBN: 978-1-53610-131-7
Editors: Adam Kovács and Patrik Nagy © 2016 Nova Science Publishers, Inc.

Chapter 5

BIOMARKER RESPONSE IN BLACK SEA SCORPION FISH *SCORPAENA PORCUS* TO ANTHROPOGENIC IMPACT

I. I. Rudneva*, E. N. Skuratovskaya, I. I. Chesnokova, V. G. Shaida and T. B. Kovyrshina
Kovalevsky Institute of Marine Biological Research RAS,
Sevastopol, Russia

ABSTRACT

Long-term and large-scale monitoring studies indicate the changes of anthropogenic impact on the water ecosystems, which can be chronically stressed by multiple environmental factors. Fish are very sensitive to anthropogenic pollution, and some of them may be tested as biomonitors for the evaluation of the ecological status and risk assessment of marine environments. Fish biochemical parameters could be directly related to the area where the fish were collected. In polluted sites the exposure of fish to xenobiotics results to interaction between these chemicals and biological systems which may give elevation to biochemical and physiological damage or/and adaptive mechanisms via the induction of defense immune and antioxidant systems. Biochemical and physiological parameters are used as biomarkers for contaminants and could be applied

*Corresponding author: I.I.Rudneva. Kovalevsky Institute of Marine Biological Research RAS, Nahimov av., 2, Sevastopol, Russia. E-mail: svg-41@mail.ru.

for the evaluation of environmental stress and its after-effects in fish. Biomarkers exposure to environmental stressors vary widely depending on the type of anthropogenic activity involved. Sewage and chemical pollution from industrial, agricultural, maritime transport and domestic effluents are the main sources of pollution in Black Sea ecosystem. Coastal waters are the main recipients of discharges and combine effects of the various kinds of contaminants result dramatic ecological consequences such as eutrophication, biodiversity loss, elimination of some species, worsening of their health and decline the population size. Biochemical biomarkers variables seem to be useful indicators of the health status of fish and their habitats in monitoring studies. Scorpion fish (*Scorpaena porcus*) is among the most common fish species in Black Sea coastal waters, and it was selected as biomonitor species. The biomarker response in scorpion fish caught in several Sevastopol bays characterizing different levels of pollution was studied related to anthropogenic impact. Selected biomarkers of exposure and effect, including parameters of oxidative stress and oxidative damage, were good tools for the evaluation of fish health and ecological status of their habitats.

Keywords: Black Sea, bays, pollution, biomarkers, oxidative stress, monitoring

1. INTRODUCTION

Marine environment and especially the coastal waters play an important role in human activity, because more than 2000000000 people live in the coastal part of the sea and ocean and use their resources directly (The World Health Report, 2002). However, despite marine coastal ecosystems are extremely productive areas, they are impacted by great anthropogenic pressure, namely from pollutants, containing in domestic, agricultural, maritime transport and industrial effluents, entering marine water. Aquatic organisms, and especially fish are very important for commercial purposes, including fishery and aquaculture. Additionally, they play the key role in ecosystem, because they belong to the vertebrates, which are at the top of the food chain. Fish health characteristics, and their responses to unfavorable living conditions, have been extensively used as indicators of habitat quality in assessment tools developed for scientific and management purposes.

Fishes' response to environmental changes is complex, depending on the intricate relations between environmental conditions and the individual ability to adapt, which depends on their life history, age, physiological status, taxonomic position (Martinez-Alvarez et al., 2005; Fonseca et al., 2011a). Usually the investigators report the elevated levels of the man-made pollution and its deleterious effects on marine organisms in all levels of their biological organization. Long-term and large-scale monitoring studies indicate the changes of anthropogenic impact on the water ecosystems, which can be chronically stressed by multiple environmental factors. Environmental pollutants transfer to fish organism and accumulate in it, caused early biological effects, especially in cell and molecular levels. Accumulation of these alterations provokes damage of organ structure, tissues and systems disfunction, biochemical changes, including ROS generation (Gillis, 2012; Gillis et al., 2014; Grinevicius et al., 2009; Kaptaner et al., 2014).

For analysis of negative factors effects on fish many different criteria are used. Fish behavior, morphological characteristics, physiological and biochemical responses to specific and multiple stressors have been extensively used to determine individual health and population status, and to assess habitat quality (Fonseca et al., 2011b; Serafim et al., 2012). A multi-biomarker approach was applied in a dynamic environment with a significant chemical contamination. Moreover, biochemical responses in natural populations of different fish species depend on their ecological and biological features, which can result from different physiological processes in them (Solé et al. 2009; Rudneva, 2012).

The present multi-biomarker approach identified of anthropogenic pollution, including industry, agriculture, fishery, aquaculture, tourism, petroleum and gas production, coastal domestic infrastructure, maritime transport impact on marine environment in Black Sea. The specificity of Sevastopol bays including their geographical position, geomorphological structure, hydrological fluctuations and different level of pollution allow to understand the main trends of anthropogenic impact on fish health and to establish the biomarkers response on stress living conditions.

In the present work the level of several biomarkers was assessed in highly distributed in Black Sea fish species *Scorpaena porcus*. For assessment of multiple biomarker responses in fish we determined the biomarkers of exposure: (1) antioxidants level in the form of enzymes activity, that reflects the response on the toxicity of reactive oxygen species (ROS):superoxide dismutase (SOD), catalase (CAT), peroxidase (PER), glutathione reductase (GR); (2) biotransformation enzyme, phase II glutathione S-transferase (GST), which metabolizes and conjugates xenobiotics; effect biomarkers: (1) lipid peroxidation (LPO) in the form of TBARS and chemiluminescence parameters; (2) level of oxidized proteins, measured as carbonyl groups concentration; (3) aminotransferase activity, which attributed with liver function. These biochemical responses were determined in scorpion fish *Scorpaena porcus*, sampled in four Sevastopol bays. It is resident species and it is also abundant in numerous other bays and coastal waters in Black Sea coast line.

The main objective of the present study was to evaluate the potential of fish biochemical response as indicators of habitat quality in marine bays, considering both natural and anthropogenic stressors.

2. Materials and Methods

2.1. Materials

Scorpion fish *Scorpaena porcus* is highly distributed benthic species in Black Sea (Figure 1). Because this fish is among the most common fish species in Black Sea coastal waters, it was selected as biomonitor organism. It is demersal form, inhabiting benthic biotops, its abundance is at the depth of 1-40 m. However, sometimes it is found at the depth of 1-10 m. Scorpion fish is a settled form, however, it migrates at the bottom of the sea and its activity is high at night. Spawning time begins at the end of May and its finish is at the middle of September. Female spawns the eggs in twin sacks, which dissolve in marine water, and the embryos in the eggs develop in pelagic zone. Scorpion fish is a batch spawner, individual female spawns many portions and the eggs hatch every 1-2 day. Individual spawning portion contains 1100-27200 eggs (Svetovidov, 1964).

Figure 1. Scorpion fish *Scorpaena porcus*.

Figure 2. Sampling sites of *Scorpaena porcus* in Sevastopol bays, Black Sea.

Fish were caught in spring period of 2014 in four Sevastopol bays Streletskaya Bay (n = 17), Aleksandrovskaya Bay (n = 18), Karantynnaya Bay (n = 20) and Kazach'ya Bay (n =10), characterizing different level of anthropogenic impact and pollution (Figure 2). The animals were immediately placed in the aerated tank, transfer in the laboratory and anesthesy. For biochemical determinations liver samples were washed several times by cold 0.85% solution of NaCl, homogenized and centrifuged at 8000 g during 15 min at cool conditions. In obtained extracts biomarkers were determined.

2.2. Biochemical Assays

Antioxidant Enzyme Activities

Antioxidant enzyme activities in the liver from scorpion fish in this study were determined according to methods described previously (Rudneva, 2012), with a few minor modifications. *Superoxide dismutase* (SOD) was assayed on the basis of inhibition of the reduction of nitroblue tetrasolium (NBT) with NADH mediated by phenazine methosulfate (PMS) under basic conditions (Nishikimi et al., 1972).All measurements were performed in 0.017 M sodium pyrophosphate buffer pH 8.3 at +25°C. The reaction mixtures contained 5 μM NBT, 78 μM NADH, 3.1 μM PMS, and a 0.1 ml sample; the final volume was

1.5 ml. The reaction was carried out in the cuvette of spectrophotometer at 560 nm. *Catalase* (CAT) was measured by the method involving the reaction of hydroperoxide reduction. *Peroxidase* (PER) activity was detected by spectrophotometric method using benzidine reagent (Litvin, 1981). The reaction mixture contained 1 ml acetate buffer pH 5,4, 0.4 ml 0.09% benzidine, 0.2 ml 0.03% H_2O_2, and 0.2 ml sample. The reaction followed in a spectrophotometer for 1 min at 20°C and at 600 nm. *Glutathione reductase* (GR) activity was assayed spectrophotometrically using a method modified after Goldberg and Sparner (1987). The reaction mixture contained 0.1 ml mM NADPH, 0.5 ml 7.5 mM oxidized glutathione, 0.2 ml mM EDTA, and 2 ml 0.05 M phosphate buffer pH 8.0. After incubation for 10 min, the extinction of the mixture was determined at 340 nm. *Glutathione-S-transferase* (GST) activity was determined by the method of Habig et al. (1974) by following the increase in absorbance at 340 nm due to the formation of the conjugate 1-chloro-2,4-dinitrobenzene (CDNB) using as substrate at the presence of reduced glutathione (GSH). The reaction mixture was prepared by mixing 1.5 ml sodium phosphate buffer 0.1 M pH 6.5, 0.2 ml GSH 9.2 mM, 0.02 ml CDNB 0.1 M and 0.1 ml of the sample. The absorbance was measured at 340 nm and at the temperature +25°C spectrophotometrically using Specol-211 (Germany). The increase in absorbance was recorded for a total 3 min. The reaction solution without the hepatic homogenates was used as blank.

Aminotransferases Activity

Aminotransferases (ALT and AST) activity was also determined spectrophotometrically with 2,4-dinitrophenylhydrazine using the standard kit (Filicit - Diagnosis, Ukraine). Specifically, 0.2 ml substrate-buffer solution was added to 0.04 ml of liver extract or serum and incubated at room temperature for 1 h. The reaction was stopped by 0.2 ml 1,4-dinitrophenylhydrazine and the solution was incubated for 20 min. Then 2 ml 0.4 N NaOH was added to the mixture and the OD of the sample was measured at 500-530 nm. The ratio AST/ALT (de Rytis coefficient) was calculated also.

Cholinesterase (ChE) Activity

Cholinesterase (ChE) was measured on the basis on hydrolysis of acetylcholinchloride to acetic acid and choline. The alterations of pH changes the colour of the mixture which was determined spectrophotometrically used the standard kit (Felicit - Diagnosis, Ukraine).

Oxidized Proteins (OP) Concentration

The analysis of the concentration of was processed according the method of Dubinina et al. (1995). Optical density of carbonyl groups reacted with 2,4-dinitrophenil hydrazones was assayed at the following wave length 346, 370, 430, 530 nm used the spectrophotometer Specol-211 (Carl Zeiss, Iena, Germany).

TBARS Level

In obtained extracts thiobarbituric acid (TBA)-reactive substances concentration was determined spectrophotometrically, according the standard procedure (Stalnaya and Garishvily, 1977), the optical density was measured at 535 nm, as we described previously (Rudneva et al., 2012).

Chemiluminescence Analysis

Parameters of spontaneous hemiluminescence (SChL) and inducible by H_2O_2 and $FeSO_4$ chemiluminescence (IChL) were measured used Luminometer 2010 (LKB, Sweden) (Vladimirov, 2001). The value of hemiluminescence was estimated in arbitrary units per protein concentration.

Total Soluble Protein Concentration

Total soluble protein concentration was quantified spectrophotometrically used the kits of the Filicit-Diagnosis (Ukraine). The enzyme activities were calculated per mg protein in the liver extracts.

2.3. Statistical Analysis

Biochemical measurements were detected in duplicate for each sample. Simple, descriptive statistics were performed using an ANOVA to determine means (+/- SEM) (Halafyan, 2008). *P* value of 0.05 was used for determination of statistical significance in all cases. The graphs were made using the computer program EXELL. Statistical correlations between studied biochemical parameters in tested animals and tissues were calculated by the least-squares method using the computer program CURVEFIT (Version 2.10-L).

Figure 3. Streletskaya Bay.

3. RESULTS

3.1. Ecological Characteristics of Tested Sites

Tested Sevastopol bays have significant anthropogenic influence from multiple sources (industrial, agricultural, shipping activities, domestic, recreation), which was evident from sediment chemical characterization. The bays are differed each from other by hydrological characteristics, hydrochemical parameters and the level of anthropogenic impact and pollution.

For the purpose of the study of pollution effects on *Scorpaena porcus* we collected fish from four Sevastopol bays, which were ranked from most (Streletskaya Bay) to medium (Aleksandrovskaya Bay and Karantinnaya Bay) and least impacted (Kazach'ya Bay).

Streletskaya Bay (Figure 3) is the most susceptible to human impacts, because of the intense navigation and recreation that occurs there. Fuel docks and shipyards are located on its coasts; their drainage, along with the discharges of two municipal treatment enterprises and two storm water sewer systems, considerably pollute the marine environment (Ovsyanyi et al., 2001; Sergeeva, et al., 2010). According to several authors (Kiryukhina and Mironov, 2004), water and sediments in Streletskaya Bay contain considerably greater amounts of oil, heavy metals, chlorinated hydrocarbons, nutrients and suspended matter than the other three bays. The level of petroleum hydrocarbons at the bottom sediments was estimated as 705 mg/100 g wet weight, which was in 10-fold higher than in Karantynnaya Bay (Eremeev et al., 2008). The mean concentration of chlorinated hydrocarbons and petroleum hydrocarbons in the bottom sediments was estimated as 1,33 и 715,5 mg/100 g wet weight (Osadchaya, 2013). To this connection the level of petroleum hydrocarbons elevated during of 2003-2009 in 1,2-1,5-fold as compared to 1990s (Osadchaya, 2013). Concentration of 6 indicative congeners of PCBs ($\sum PCB_6$) and $\sum$ DDT in bottom sediments in Streletskaya Bay was estimated as 121 and 51 ng/g respectively (Malahova and Malahova, 2009), while mercury level was higher in 5-10-fold than the legal levels and ranged from 460 to 1063 ng/g dry weight (Kostova and Plotitcina, 2011). However, in the recent decade a change of the "quality" of the anthropogenic load was recorded in Streletskaya Bay, due to intensive building of tourist hotel complexes and shipyards on its coast.

Aleksandrovskaya Bay is involved in the great Sevastopol Bay and it is located before the pier, which separates Sevastopol Bay from the open part of the sea (Figure 4). The sediments of Aleksandrovskaya Bay were contaminated to a greater extent and they classify as the third level of oil contamination (Rubtzova, et al., 2013). Level of $\sum PCB_6$ и $\sum$ DDT in the bottom sediments in this bay was estimated as 97,4 ng/g dry weight and 14,8 ng/g dry weight respectively (Shcurova and Malahova, 2012). At the period of 1985 to 2009 concentration of petroleum hydrocarbons (PAH) in bottom sediments tended to increase in Sevastopol Bay, including in Aleksandrovskaya Bay also, which is located at the entrance of Sevastopol Bay (Osadchaya, 2013).

Figure 4. Aleksandrovskaya Bay (1) and Karantinnaya Bay (2).

Karantinnaya Bay is located near Sevastopol Bay. They are separate each from other by the pier (Figure 4). In this bay the level of chlorinated hydrocarbons and petroleum hydrocarbons in the bottom sediments is estimated as 165,0 и 65,0 mg/100 g wet weight respectively (Eremeev et al., 2008). However, water in this bay contains NO_3^- 17-116μg/l and $N_{org.}$ 552 μg/l, which was greater than in the other Sevastopol bays (Kuftarkova et al., 2008). That was due to the different locations of the areas and the discharges of domestic wastewater in Karantinnaya Bay.

Kazach'ya Bay is located at 15 km from the center of the city Sevastopol between the Kamyshovaya Bay and the cape Phiolent (Figure 5). Kazach'ya Bay is selected as reference site in the systems of Sevastopol bays, because the level of toxicants in its sediments is less, than in other sampling sites. For instance, the level of mercury was estimated as 33.7 ng/g dry weight and PCB content was 152.0 ng/g wet weight (Malahova et al., 2003). Concentration of chlorinated hydrocarbons and petroleum hydrocarbons was estimated as 157.0 mg/100 g and 36.0/mg/100g (Eremeev et al., 2008) respectively, which is the evidence of its low polluted status as compared with the other tested sites (Rubtzova et al., 2013). According the data of the chemical analysis, concentration of NO_2^- in the water was estimated as 1,0 μ/L, NO_3^- 6 μ/L, NH_4^- 11-14 μ/L, and $N_{org.}$ 447 μ/L, that is the evidence of the least anthropogenic press on the ecosystem of Kazach'ya Bay as compared to the other tested sites (Kuftarkova et al., 2008).

Figure 5. Kazach'ya Bay.

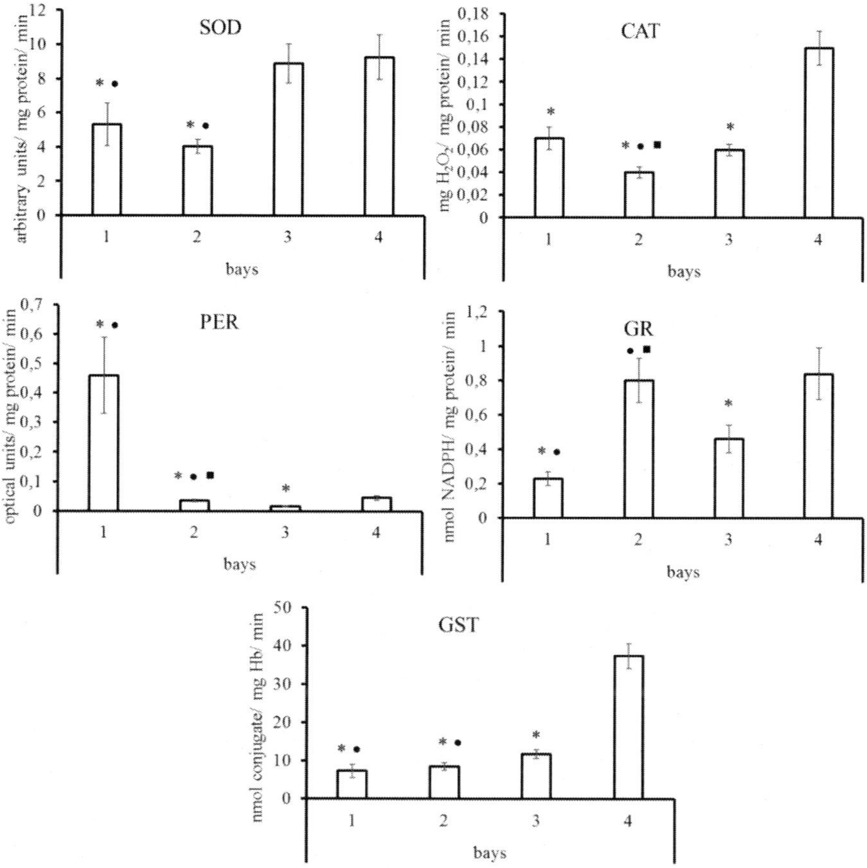

Figure 6. Antioxidant enzyme activities in the liver of scorpion fish, mean + SEM (1 – Kazach'ya Bay, 2 – Karantinnaya Bay, 3 – Aleksandrovskaya Bay, 4 – Streletskaya Bay; * – significant differences between fish from Streletskaya Bay and the values in fish from another bays; • – with Aleksandrovskaya Bay, ▪ – with Kazach'ya Bay).

Despite of the additional non-persistent source of pollution, which is dolphinarium, located in the bay, according the chemical characteristics of the water and sediments, we suggest that Kazach'ya Bay can be used as reference area in comparative studies of biomarkers in scorpion fish for the evaluation of pollution effects caused anthropogenic activity. Moreover, the processes of the self-cleaning in this site are very intensive, which provides to save natural environment in general (Moiseenko, 2012).

Therefore, based on the analysis of the published data, we can conclude that among the tested Sevastopol bays Streletskaya Bay is the most negatively affected environmentally, Karantinnaya Bay and Aleksandrovskaya Bay are less contaminated, and Kazach'ya Bay is least pressed. To the other hand, according the data of Kiryukhina and Mironov (2004), sediments in Aleksandrovskaya Bay were contaminated to a greater extent than those in Karantinnaya Bay, and the surface layers of the water, in contrast, were not polluted. That was due to the different locations of the areas and the discharges of domestic wastewater in Karantinnaya Bay.

3.2. Biomarkers Response

Antioxidant Enzyme Activity

High SOD activity ($p<0.05$)was indicated in the liver of fish from Streletskaya Bay and Aleksandrovskaya Bay in comparison to the values of the individuals from the other sampling sites (Figure 6). No significant differences were found in SOD activity in fishes from Karantinnaya Bay and Kazach'ya Bay. Enzymatic level was the similar in the animals, caught in Streletskaya Bay and Aleksandrovskaya Bay. The greatest CAT activity ($p<0.05$) was measured in the fish from Streletskaya Bay. The lowest activity ($p<0.05$) was observed in the animals from Karantinnaya Bay, while no significant differences were shown between the CAT level in scorpion fish inhabiting Aleksandrovskaya Bay and Kazach'ya Bay. PER activity was more, than in 10-fold greater ($p<0.01$)in the liver of fish from Kazach'ya Bay, the lowest value was shown in fish caught in Aleksandrovskaya Bay, no significant differences were indicated in the animals from Karantinnaya Bay and Streletskaya Bay. High GR activity was shown in the fish from Streletskaya Bay and Karantinnaya Bay, the lowest value was indicated in the animals from Kazach'ya Bay, while GR activity in the fish from Aleksandrovskaya Bay was greater in 2-fold ($p<0.05$). The highest GST activity was shown in the individuals from Streletskaya Bay ($p<0.05$), enzyme activity was the similar in the fish from Kazach'ya Bay and Karantinnaya Bay, while the value of the individuals from Aleksandrovskaya Bay was higher ($p<0.05$). No relationships were observed between antioxidant enzyme activities in tested fish.

Cholinesterase Activity

ChE activity was significant higher ($p<0.05$) in the liver of fish from Kazach'ya Bay as compared to the values of the animals from the other tested bays (Figure 7). No significant differences were observed in ChE level in the fish from Streletskaya Bay, Karantinnaya Bay, and Aleksandrovskaya Bay.

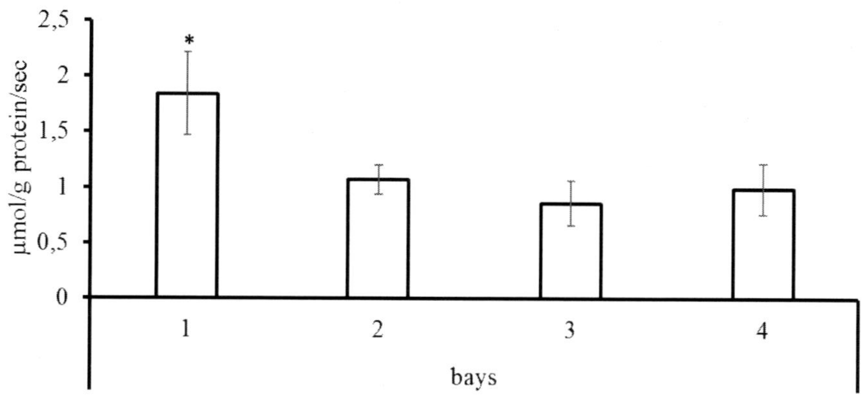

Figure 7. Cholinesterase activity in the liver of scorpion fish, mean ± SEM, 1-4 - bays, see Figure6 (* - significant ($p<0.05$) difference with the values of the fish captured in the tested bays).

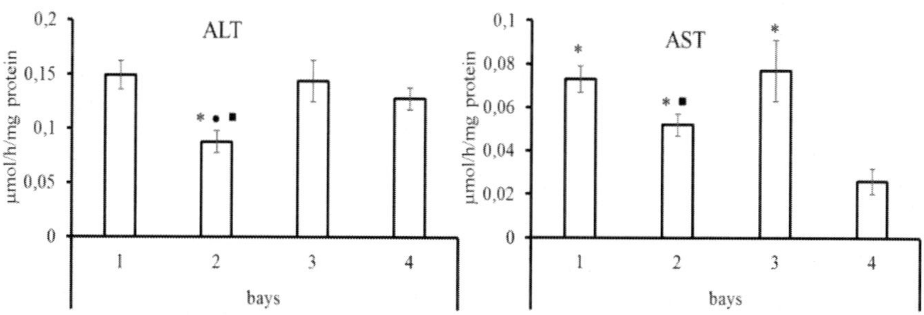

Figure 8. Aminotransferase activity in the liver of scorpion fish, mean + SEM, 1 – 4 bays, see Figure 6. * – significant differences between the fish from Streletskaya Bay and the values in fish from another bays; • – with Aleksandrovskaya Bay, ▪ – with Kazach'ya Bay).

Aminotransferase Activity

Aminotransferase activity is present in Figure 8. The lowest ALT activity was indicated in the fish from Karantinnaya Bay ($p < 0.05$), while no differences were observed between enzyme activities in the liver of the fish from the other tested bays. High AST activity was shown in the scorpion fish caught in Kazach'ya Bay and Aleksandrovskaya Bay ($p < 0.05$), the lowest value was measured in the fish from Streletskaya Bay, hepatic AST activity in the individuals from Karantinnaya Bay was intermediate as compared with the values of the fish from the other tested sites.

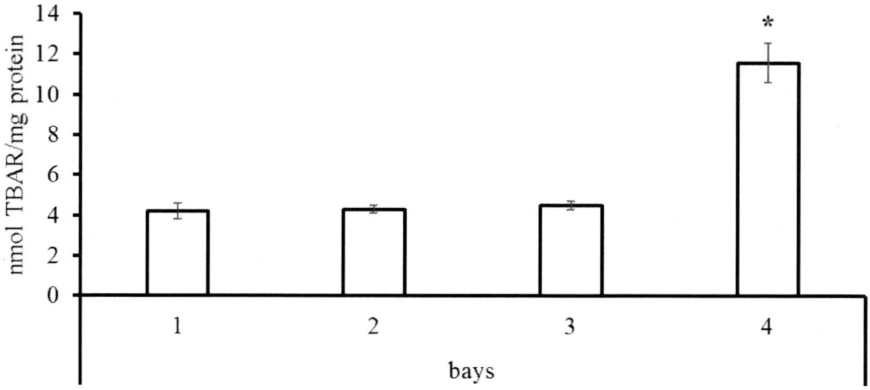

Figure 9. Concentration of TBA-reactive substances in the liver of scorpion fish, mean + SEM. 1-4 bays, see Figure 6. * - significant ($p < 0.05$) difference with values of the fish captured in the tested bays).

TBARS Concentration

The greatest TBARS level was indicated in the liver of fish, collected in Streletskaya Bay ($p < 0.01$), while no significant differences were observed between the values of the individuals from the other examined bays (Figure 9).

Chemiluminescence Level

The values of spontaneous chemiluminescence (SChL) level was significantly greater ($p<0.05$) in the hepatic extracts of the fish, caught in Streletskaya Bay and Aleksandrovskaya Bay, as compared with the values of the individuals from the Kazach'ya Bay and Karantinnaya Bay (Figure 10). The response of inducible chemiluminescence (IChL) varied widely in hepatic extracts of fish from tested bays: it was the highest in the fish from Streletskaya Bay and progressively decreased in the individuals as following: Kazach'ya Bay→ Karantinnaya Bay→Aleksandrovskaya Bay.

Oxidized Proteins (OP) Concentration

Concentration of oxidized proteins (OP) in the liver of scorpion fish from checked bays is present in the Figure 11. Level of oxidized proteins, measured in all tested wave length (OD_{346}, OD_{370}, OD_{430}, and OD_{530}) was the greatest in the fish from Streletskaya Bay, high values were shown in the fish from

Karantinnaya Bay and Aleksandrovskaya Bay, and the lowest level was found in the individuals caught in Kazach'ya Bay.

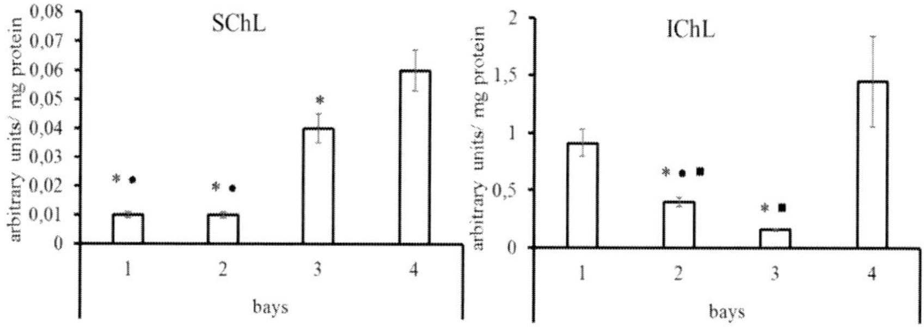

Figure 10. Chemiluminescence level in the liver of scorpion fish, mean + SEM. SChL – spontaneous, IChL – inducible. 1 – 4 – bays, see Figure 6. * – significant differences between the fish from Streletskaya Bay and the values in fish from another bays; • – with Aleksandrovskaya Bay, ▪ – with Kazach'ya Bay.

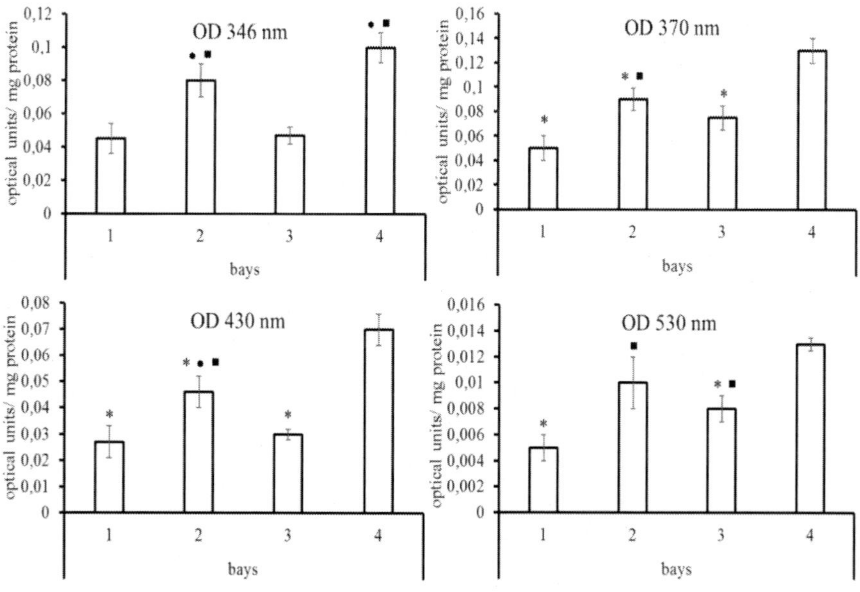

Figure 11. Concentration of oxidized proteins in the liver of scorpion fish, mean + SEM.1-4 bays, see Figure 6. * – significant differences between fish from Streletskaya Bay and the values in fish from another bays; • –with Aleksandrovskaya Bay, ▪ – with Kazach'ya Bay.

Therefore, we conclude the different response in the hepatic biomarkers in the scorpion fish, caught in examined Sevastopol bays.

4. Discussion

The ecological characterization of examined bays in the coastal part of Sevastopol city showed the presence of diverse contaminants in the sediments and in the water, which are the result of human activities. The main anthropogenic sources of these pollutants are attributed with industrial and agricultural discharges, domestic effluents, shipping and recreation activities. High concentration of xenobiotics in the bottom sediments and in the water in the biotops of scorpion fish causes oxidative stress in fish resulted their transfer via water and food in the organism.

In polluted areas the exposure of aquatic organisms to xenobiotics results to interaction between these compounds and biological systems which may give elevation to biochemical and physiological damage or/and adaptive mechanisms via the induction of defense immune and antioxidant systems (Goksoyr et al., 1996). Therefore, biochemical and physiological characteristics are used as biomarkers for contaminants and could be applied for evaluation of environmental stress and its after-effects. However, biomarkers exposure to environmental stressors vary widely depending on the type of anthropogenic activity involved (Adams, 2005). Biomarkers response depends on tissues and organs specificity and functions. In this context, liver is the main organ of pollutants accumulation and biotransformation, generating ROS production and its metabolites (Sole et al., 2009; Grinevicius et al., 2009; He et al., 2012; Morachis-Valdez et al., 2015).

For the evaluation of oxidative stress in fish caught in different bays, we used various parameters, and among them we studied the biomarkers of exposure - the activities of antioxidant enzymes namely SOD, CAT, PER, GR and GST, which was also recognized as phase II biotransformation enzyme (Livingstone, 2001; Van der Oost et al., 2003). Enzymatic antioxidant response in the liver of scorpion fish was characterized by a higher variability of SOD and GR activities among tested bays, while CAT, PER and GST values were more constant. In this case our results agree with the data of other researchers, who also reported the similar trends of antioxidant enzymatic response in the liver of *Dicentrarchus labrax,* caught in the estuaries, impacted by human activity (Fonseca et al., 2011a).

Antioxidant enzyme activity in the liver of fish from the most contaminated Streletskaya Bay was significantly higher as compared with the values of the animals, caught in other tested sites, with the exception of PER. Furthermore, antioxidant enzyme activity had positive correlation (with the exception of PER) with the level of anthropogenic impact in the examined bays, suggesting that pollution level increases the ROS production in fish liver and the antioxidant enzymatic system responds to an increase of environmental stress. To the other hand, increase of antioxidant enzyme activities in the fish from the most polluted bay demonstrated their ability to resist against oxidative stress and to adapt to the unfavorable living conditions. The similar responses of enzyme antioxidant system were documented in the tissues of fish affected different levels of pollutants in the laboratory and field conditions: heavy metals (Cao et al., 2010, 2011), petroleum and oil compounds resulting oil spills (Martinez-Gomez et al., 2009), organic chemicals (Al-Mamoori et al., 2014), complex urban effluents (Hansson et al., 2006), etc. We could note the specificity response of antioxidant enzymes in fish liver to different toxicants and their mixture, containing in the water and sediments as well as synergic and antagonistic effects at the case of interactions between xenobiotics and water. However, extremely high level of hepatic PER activity in scorpion fish from free contaminated Kazach'ya Bay could be attributed with the presence short term local pollution source, which induced enzyme activity in the hepatocytes of the animals in the least impacted site.

GST activity was significantly higher (approximately in 4.5-5-fold) in the liver of fish from the most impacted Streletskaya Bay, while no significant differences were observed among the values of the individuals from less polluted and reference sites. We also noted high positive correlation between CAT and GST activity ($r=0.87$), while no relationships were observed between the other examined antioxidant enzyme values. GST plays an important role in the detoxification of lipid peroxides and demonstrates the functions such as glutathione peroxidase activity towards reactive oxygen species in the cells under oxidative stress. Induction of GST activity in some aquatic organisms such as mussels has been also found in high polluted marine environments (Hansson et al., 2006), after the wreckage and oil spills of the tankers (Martinez-Gomez et al., 2009). Increase of GST activity was documented in some fish species and invertebrates collected in environments impacted by complex discharges of contaminants and accidents (Hamed et al., 2004). Because enzyme plays an important role in biotransformation of xenobiotics, it may be used as a valid bioindicator for evaluation of pollution level. GST clearly responds on the pollution level and its response is linked with the antioxidant enzyme activities and induction of biotransformation processes in the hepatocytes. At the other hand, different factors both biotic, abiotic and anthropogenic may be determined GST activity and in our previous study we described the fluctuations of GST level in several fish species, inhabiting different biotops in the coastal waters of Black Sea (Rudneva et al., 2010). The researchers reported also, that the spatial patterns in GST activity were complex, given that high enzyme activity was observed in the liver of teleost species in highly polluted sites and in the least polluted site. The investigators suggest a strong GST inhibition by organic compounds in the liver of two species *S. senegalensis* and *P. microps* from the site, where the highest PAH concentrations were measured (Fonseca et al., 2011b).

Oxidative damage in scorpion fish liver measured as effect biomarkers, namely TBARS, oxidized proteins and chemiliuminescence level both spontaneous and inducible, showed the variability for between tested bays. Our findings indicated significant increasing (more than in 2-fold) the level of TBARS in the liver of scorpion fish caught in the most polluted Streletskaya Bay, than the corresponding values of the animals from the less polluted sites, which were comparable. High positive correlation was observed between CAT activity and TBARS level (r=0.89) and GST activity and TBARS (r=0.98). Therefore, we suggest, that TBARS level is a good biomarker for the evaluation of fish response to pollution. The researchers also reported the increase of LPO values in the aquatic animals exposed to pollutants (Javed et al., 2015; Jiang et al., 2011).

Similar trends were shown in chemiluminescence levels, which were attributed with the increasing of lipid peroxidation in the liver of fish, inhabiting polluted environment (Vladimirov, 2001). Both IChl and SChL values were significantly greater in the liver of the animals from the most impacted Streletskaya Bay. We found high positive correlation between ChL values in fish hepatocytes and TBARS concentration (r=0.51) and between ChL level and CAT activity (r=0.67). However, we measured high IChL level in the fish from the reference site Kazach'ya Bay, which could be explained for the complex influence of biotic and abiotic factors on the ROS production. Different food spectra, hydrochemical and hydrological fluctuations may be the reasons of high level of inducible chemilumioniscence in the liver of scorpion fish from Kazach'ya Bay, while the spontaneous (SChL) was lower as compared with the values of the fish from less impacted sites Aleksandrovskaya Bay and especially Streletskaya Bay. Therefore, we could conclude that ChL level in hepatocytes is also good biomarker in the procedure of the evaluation of fish health in examined sites.

Concentration of oxidized proteins (OP) in tissues is an indicator of the chronic pollution impact on fish organism resulted oxidative stress also. Oxidative stress caused unfavorable living conditions is an imbalance toward the pro-oxidant side of the pro-oxidant/antioxidant homeostasis. The content of protein carbonyl groups are good biomarker of oxidative stress because they have some advantages in comparison with the measurement of other oxidative products: they are the relative early formation and the relative stability of carbonylated (oxidized) proteins (Dalle-Donne et al., 2003; Luschak, 2007). Carbonyl (CO) groups (aldehydes and ketones) are produced on protein side chains when they are oxidized. These moieties are chemically stable, therefore protein carbonyl content is the most general indicator of protein oxidation. The accumulation and increase of carbonyls in proteins has been observed in several human diseases (Dalle-Donne et al., 2003; Luschak, 2007). Oxidative damage of proteins may result negative consequences because among proteins there are enzymes, hormones, mediators and other biomolecules, which play an important role in the metabolism and homeostasis of the organism.

In our studies we showed significant increase of oxidized proteins concentration in the liver of fish from the most contaminated Streletskaya Bay. The total level of oxidized proteins increased progressively in the hepatocytes of the individuals from Kazach'ya Bay→Karantinnaya Bay→Aleksandrovskaya Bay→Streletskaya Bay (0.127→0.226→0.16→0.313 correspondingly). We found high direct correlation between TBARS level and oxidized proteins concentration (r= 0.63). However, high correlations between GR, GT and OP were shown also (r=0.76 and r=0.61 respectively), which could reflect the interactions between the antioxidant enzyme activities and OP production. Growth of LPO parameters and oxidized proteins level in tissues of the animals in contaminated habitats are postulated as the non-specific response of the organism to the stress effects, and it was documented at the case of environment pollution by various chemical agents and their complex (Martinez-Porchas et al., 2011; Liu et al., 2006; Kaptaner et al., 2014).

Oppositely, we found that hepatic ChE activity progressively decreased in fish as following Kazach'ya Bay→ Karantinnaya Bay→Aleksandrovskaya Bay→Streletskaya Bay. Inhibition of enzyme activity may reflect the pollution of biotope of organophosphate chemicals, metals, detergents which were recognized in domestic sewage and maritime effluents entering in Streletskaya Bay. Our results agree with the data of several researchers, who documented the inhibition of ChE activity in tissues of fish in polluted locations (Van der

Oost et al., 2003; Napierska et al., 2009). ChE inhibition by organophosphorus (OP) compounds follows different behaviors depending on pesticide chemical structure. Ions can alter ChE activity inhibiting or activating so that it was proposed that enzymes could be used as biomarkers of heavy metals pollution. Another concern about application fish ChEs as biomarker of pesticides is that in eutrophed sites some species of cyanobacteria produce anticholinesterasic metabolites such as anatoxin-a which can be considered natural OP compounds and which toxicity can be approximately in 1000-fold higher than that of the insecticide paraoxon (Assis et al., 2011). Cholinesterases may reflect the disturbance of neural functions of the organism in stress conditions. Monitoring of ChE inhibition has been measured in the tissues of marine fish seems to be a valid indicator for chemical pollution of the marine environment as well (Kirbi et al., 2000). In our previous study we described the differences of ChE activity in the blood of scorpion fish caught in different polluted areas and we also showed significantly decrease of the enzymatic activity in the blood serum of the individuals in the most impacted site (Rudneva et al., 2012).

AST is used as clinic diagnostic tool and it is attributed with cell necrosis of the liver and skeletal or cardiac muscle, starvation and lacking of vitamin E. ALT is an acute hepatic damage good marker (Coppo et al., 2001-2002). Hepatic ALT activity in scorpion fish varied less, than AST. The lowest ALT activity was indicated in the fish in Karantinnaya Bay without further site variability. AST activity varied independently and the lowest value was shown in the animals in the most polluted Streletskaya Bay. Both ALT and AST activities were lower in fish caught in Karantinnaya Bay. Therefore, at present study we reported significantly decrease of AST activity in the liver of fish in Streletskaya Bay, which can be attributed with the great intoxication, caused accumulation in the organ high concentration of toxicants. Enzyme activity could be inhibited by the complex mixture of environmental contaminants in Streletskaya Bay. In our previous study we also reported that aminotransferase activities (both ALT and AST) were lower in blood serum of fish in polluted sites (Streletskaya Bay and Karantinnaya Bay) than those in the samples collected in free polluted areas (Rudneva et al., 2012). We propose that the reason of this phenomena was attributed with the damaged of hepatocytes in the liver of fish in highly impacted sites. To the other hand, aminotransferases are the organ-specific indicators for toxic effects and determination of transaminases in the liver has proved in diagnosis of liver damage (Van der Oost et al., 2003). Thus, the response of hepatic transaminases is not uniform and could be used as indicator of fish physiological status and health.

Therefore, we can conclude, that tested biochemical parameters in scorpion fish are good biomarkers for the evaluation of oxidative stress, caused anthropogenic pollution. However, they respond clearly at the case of the strong differences in pollution level between examined sites. Oppositely, the response was not uniform in fish from both less polluted bays and, probably, it depends on various xenobiotic combinations in them. Additionally, in these sites we could have so-called “mask effect” of pollutants which was associated with the complex effects of pollutants mixture on fish enzymatic activities. Some of toxicants induce oxidative stress and increase the level of antioxidant enzyme activities, while the others (heavy metals) bind with protein amino acids, SH-groups, carbonyls, and other reactive groups, damage enzyme structure and inhibit its activity. We observed this effect in fish blood in highly polluted area, where lack or so-called “normal” enzyme activities were the result of summary effects of combinations of pollutants and their interactions (Rudneva et al., 2012). These interactions could modulate organism response and lead not only adaptive effect, but toxic one, characterizing the inhibition of defense systems, including immune and antioxidant components. Thus, in some cases in the tissues of fish inhabiting high polluted water bodies the researchers observed the decrease of the activity of antioxidants and other key enzymes, which characterized clearly toxic effect caused high level of heavy metals and chlororganic compounds accumulation.

CONCLUSION

Biomarkers variables in the liver of scorpion fish from different sites of Sevastopol coastal waters, reported in this study, seem to be useful indicators of the health status of fish and water quality in monitoring studies. The best biomarkers are the following: ChE activity and GST activity, TBARS, oxidized proteins and chemiluminescence level, which are highly sensitive to anthropogenic pollution and reflect the response of the organism on stressful environment. Antioxidant enzyme and aminotransferase activities response was not uniform and depends on examined site and the combinations of pollutants in them. The obtained results can be applied for development monitoring management and for perspectives of conservation ecology and fish status in impacted aquatic ecosystems. The analysis of the hepatic biomarkers is important for the evaluation of fish health in human impacted water bodies and their abilities to protect against chemical pollution and keep their life in the pollute environments. As the response of biomarkers are not uniform and only measuring a battery of bioindicators can provide insight into the fish health and ecological status of their habitats.

Acknowledgments

The authors wish to thank fishermen of IMBR, to all involved in fish sampling and laboratorial procedures, Mr Alex Jurkov and Mr. Oleg Shaida for the pictures of the Sevastopol bays in this paper.

References

Adams, S. M. (2005). Assessing cause and effect of multiple stressors on marine systems. *Marine Pollution Bulletin*, 51 (8-12), 649-657.

Al-Mamoori, A. M. J., Al-Zubaidy, F. M., Abd Al-Rezzaq, A. J., Adil Hadi, M. and Hass, M. J. (2014). Biomarkers of Chlorofos toxicity in Common Carp *Cyprinus carpio.IOSR-JESTFT*, 1, 109-112.

Assis, C. R. D., Bezerra, R. S. and Carvalho, L. B. (2011). Fish cholinesterases as biomarkers of organophosphorus and carbamate pesticides. In: M. Stoytcheva (Ed.), *Pesticides in the Modern World – Pests Control and pesticides Exposure and Toxicity Assessment* (pp. 254-278). US: InTech.

Cao, L., Huang, W., Liu, J., Yin, X. and Dou, Sh. (2010). Accumulation and oxidative stress biomarkers in Japaneese flounder larvae and juveniles under cadmium exposure. *Comp. Biochem. Physiol.*, 151 (3) C, 386-392.

Cao, L., Huang, W., Shan, X., Ye, Z. andDou, Sh. (2011). Tissue-specific accumulation of cadmium and its effects on antioxidative responses in Japanese flounder juveniles. *Environmental toxicology and pharmacology*, 33 (1), 16-25.

Coppo, J. A., Mussart, N. B. and Fioranelli, S. A. (2001-2002). Physiological variation of enzymatic activities in blood of bullfrog, *Rana catesbeianna* (Shaw, 1802). *Reviews of Veterinary*, 12/13 (1-2), 22-27.

Dalle-Donne, I., Rossi, R., Giostarini, D., Milzani, A. andColombo, R. (2003). Protein carbonyl groups as biomarkers of oxidative stress. *Clinica Chimica Acta*, 329, 23-38.

Dubinina, E. E., Burmistov, S. O., Hodov, D. A. and Porotov, I. G. (1995). Oxidative modification of human serum proteins, the method of determining. *Proceedings of medicine chemistry*, 1, 24-26. (in Russian).

Eremeev, V. N., Mironov, O. G., Alemov, S. V., Burdiayn, N. V., Shadrina, T. V., Tihonova, E. A., Volkov, N. G. andIstomina, M. I. (2008). Preliminary results of the estimation of petroleum pollution of Kerch Strait after the oil spills. *Marine Ecological J.*, 7 (3), 15-24. (*in Russian*).

Fonseca, V., França, S., Serafim, A., Company R., Lopes B., Bebianno M. J. and Cabral H. N. (2011a). Multi-biomarker responses to estuarine habitat contamination in three fishspecies: *Dicentrarchus labrax*, *Solea senegalensis* and *Pomatoschistus microps*. *Aquatic Toxicology*, 102 (3), 216-227.

Fonseca, V., França, S., Vasconcelosa, R. P., Serafim, A., Company, R., Lopes, B., Bebianno, M. J. and Cabral, H. N. (2011b). Short-term variability of multiple biomarker response in fish from estuaries: influence of environmental dynamics. *Mar. Environ. Res.*, 72 (4), 172-178.

Gillis, P. L. (2012). Cumulative impacts of urban runoff and municipal wastewater effluents on wild freshwater mussels (*Lasmigona costata*). *Sci. Total Environ.*, 431, 348-356.

Gillis, P. L., Higgins, S. K. and Jorge, M. B. (2014). Evidence of oxidative stress in wild freshwater mussels (*Lasmigona costata*) exposed to urban-derived contaminants. *Ecotoxicol. Environ. Saf.*, 102, 62-69.

Goksoyr, A., Beyer, J., Egaas, E., Grosvik, B. E., Hylland, K., Sandvik, M. and Skaare, J. U. (1996). Biomarker responses in flounder (*Platichthys flesus*) and their use in pollution monitoring. *Marine Pollution Bulletin*, 33, 36-45.

Goldberg, D. M., and Sparner, R. J. (1987). Glutathione reductase. In: H.U. Bergmeyer, J. Bergmeyer, and M. Grab (Eds). *Methods of Enzymatic Analysis*. Vol. III. Verlag Chemic (pp. 258-265). Weinheim.

Grinevicius, V. M. A. S., Geremias, R., Lank, R., Bettega, K. F., Laranjeiras, M. C., Fuvere, V. T., Filho, D. W. and Pedrosa, R.C. (2009). Textile effluents induce biomarkers of acute toxicity, oxidative stress, and genotoxicity. *Arch. Environ. Contam. Toxicol.*, 57, 307-314.

Grune, T. (2000). Oxidative stress, aging and the proteasomal system. *Biogerontology*, 1, 31-40.

Habig, W. H., Pabst, M. J. and Jokoby, W.B. (1974). Glutathione S-transferase. The first enzyme step in mercapturic acid formation. *J. Biol. Chem.*, 249, 7130-7139.

Halafyan, A. A. (2008). *Data statistic processing. Analysis of Variance. In STATISTICA 6* (pp. 133-152). Moscow: Binom. (*in Russian*).

Hamed, R. R., Maharem, M. and Guinidi, R. A. M. (2004). Glutathione and its related enzymes in the Nile Fish. *Fish Physiology and Biochemistry*, 30, 189-199.

Hansson, T., Schiedek, D., Lehtonen, K. K., Vuorinen, P. J., Liewenborg, N., Noaksson, E., Tjarnlund, U., Hansson M., and Balk, L. (2006). Biochemical biomarkers in adult female perch (*Perca fluviatilis*) in a chronically polluted gradient in the Stockholm recipient (Sweden). *Marine Pollution Bulletin*, 53, 451-468.

He, X., Nie X., Yang, Y., Liu, X., Pan, D., Cheng, Z. and Liang, X. (2012). Multi-biomarker responses in fishes from two typical marine aquaculture regions in South China. *Mar. Pollut. Bull.*, 64, 2312-2324.

Javed, M., Usmani, N., Ahmad, I., Ahmad, M. (2015). Studies on the oxidative stress and gill histopathology in *Channa punctatus* of the canal receiving heavy metal-loaded effluent of Kasimpur Thermal Power Plant. *Environmental Monitoring and Assessment*, 187 (1), 11p.

Jiang, J., Gu, X., Song, R., Zhang, Q., Geng, J., Wang, X., Yang, L. (2011). Time-dependent oxidative stress and histopathological changes in *Cyprinus carpio* L. exposed to microcystine-LR. *Ecotoxicology*, 20, 1000-1009.

Kaptaner, B., Kankaya, E., Doğan, A. and Çelik, I. (2014). Histopathology and oxidative stress in the liver of *Chalcalburnus tarichi* living in Lake Van, Turkey. *Life Sci. J.*, 11 (8), 66-77.

Kostova, S. K. and Plotitcyna, O. V. (2011). Chemoecological studies of mercury along Black Sea Crimean coastal zone Chemical education in the context of chemical security: status of the problem and the perspectives. In:*Abstract Book of Int. Sci-Practic Conf.* (pp. 64-66). Kiev: National Pedagogical University named Dragomanov. (*in Russian*).

Kirby, M.F., Morris, S., Hurst, M., Kirbi, S.J., Neall, P., Tylor, T. and Fagg, A. (2000). The use of holinesterase activity in flounder (*Platichthys flesus*) muscle tissue as a biomarker of neurotoxic contamination in UK estuaries. *Marine Pollution Bulletin*, 40 (9), 780-791.

Kiryukhina, L. N. and Mironov, O. G. (2004) Chemical and microbiological characteristics of bottom sediments of the Sevastopol Bay. *Ekologiya Morya*, 66, 53-58. (*in Russian*).

Kuftarkova, E. A., Rodionova, N. Y., Gubanov, V. I. and Bobko, N. I. (2008). Hydrochemical characteristics of several Sevastopol bays. *Proc. Of*

Southern State Scientific Research Institute of Oceanography. 46. 110-117. (*in Russian*).

Litvin, F. F. (1981). *Laboratory manual of physicochemical methods in biology*. Moscow: Moscow State University. (*in Russian*).

Liu, H., Wang, W., Zhang, J. and Wang, X. (2006). Effects of copper and its ethylenediaminetetraacetate complex on the antioxidant defenses of the goldfish, *Carassius auratus*. *Ecotoxicological Environmental Safety*, 65(3), 350-354.

Livingstone, D. R. (2001). Contaminant-stimulated reactive oxygen species production and oxidative damage in aquatic organisms. *Marine Pollution Bulletin*, 42, 656-665.

Luschak, V.I. (2007). Free radical protein oxidation and its link with organism functional status. *Biochemistry*, 72 (8), 995-1017. (*in Russian*).

Malahova, L.V., Kostova, S. K. and Plotytzina, O. V. (2003). Chemical pollution of the components in the ecosystem of Kazach'ya Bay (Black Sea). In: *Ecological Safety of Coastal and Shelf Zones of the Sea and complex application of shelf resources*, 9. (pp. 112-116). Sevastopol: EKOSI-Gidrofizika. (*in Russian).*

Malahova, L. V. and Malahova, T. B. (2009). Persistent organic pollutants in Sevastopol coastal waters and bays (Black Sea). *V Int. Conf. «Ecological problems of the great and small cities and industrial zones»*. (pp. 120-121). Saint-Petersburg: Conference Proc. Saint-Petersburg, Crismas+Saint-Petesburg. *(in Russian).*

Martinez-Alvarez, R., Morales, A. and Sanz, A. (2005). Antioxidant defenses in fish: biotic and abiotic factors. *Rev. Fish Biology and Fisheries*, 15, 75-88.

Martinez-Gomez, C., Fernandez, B., Valdes, J., Campillo, J. A., Benendicto, J., Sanchez, F. and Vethaak, A. D. (2009). Evaluation of three-year monitoring with biomarkers the Prestige oil spill (Spain). *Chemosphere*, 74 (5), 613-620.

Martinez-Porchas M., Hernandez-Rodriguez M., Davila-Ortiz J., Vila-Cruz V. and Ramos-Enriquez J.R. (2011). A preliminary study about the effect of benzo[α]pyrene (BaP) injection on the thermal behavior and plasmatic parameters of the Nile tilapia (*Oreochromis niloticus* L.) acclimated to different temperatures. *Pan-American Journal of Aquatic Sciences*, 6(1),76-85.

Moiseenko, O. G. (2012). Evaluation of the status of carbon cycle of the coastal waters of Black Sea. In: *Current fishering and ecological problems of the region of Black Sea and the Sea of Azov. Proceedings of the VII International conference*, 1. (pp. 265-272). Kerch: JugNIRO. *(in Russian).*

Morachis-Valdez, G., Dublán-García, O., López-Martínez, L. X., Galar-Martínez, M., Saucedo-Vence, K.andGómez-Oliván, L. M. (2015). Chronic exposure to pollutants in Madín Reservoir (Mexico) alters oxidative stress status and flesh quality in the common carp *Cyprinus carpio*. *Environmental Science and Pollution Research*, 22 (12), 9159-9172.

Napierska, D., Barsiene, J., Mulkiwicz, E., Podolska, M. and Rybakovas, A. (2009). Biomarker responses in flounder *Platichthys flesus* from the Polish coastal area of the Baltic Sea and applications in biomonitoring. *Ecotoxicology*, 18, 846-859.

Nishikimi, M., Rao, N.A. and Yagik, K. (1972). The occurance of superoxide anion in the reaction of reduced phenazine. *Biochem. Biophys. Res.Commun.*,46 (2), 849 -854.

Osadchaya, T. S. (2013). Petroleum hydrocarbons in the bottom sediments of the coastal waters of Sevastopol (Black Sea). Scientific research and their practical application. Modern state and ways of development 2013. http://www.sworld.com.ua/index.php/ru/conference/the-content-of-conferences/archives-of-individual-conferences/oct-2013. *(in Russian).*

Ovsyanyi, E. I., Romanov, A. S., Min'kovskaya, R. Y. et al., (2001). The main sources of pollution of the marine environment of the Sevastopol region. In:*Ecological Safety of Coastal and Shelf Zones and Comprehensive Use of Shelf Resources*, 2. (pp. 138-152). Sevastopol: EKOSI-Gidrofizika. *(in Russian).*

Rubtzova, S. I., Tihonova, E. A., Burdiyan, N. V. and Doroshenko, Y. V. (2013). Evaluation of ecological status of Sevastopol bays in Black Sea used the basic chemical and microbial criteria. *Marine ecological J.*, 12(2), 38-50. (in Russian).

Rudneva, I. I. (2012). Antioxidant defence in marine fish and its relationship to their ecological status. In: S. P. Dempsy (Ed.), *Fish Ecology*. (pp. 31-59). New York: Nova Science Publishers.

Rudneva, I. I., Kuzminova, N. S., Skuratovskaya, E. N. and Kovyrshina, T. B. (2010). Comparative study of glutathione-S-transferase activity in tissues of some Black Sea teleosts. *Int. J. Sci. Nat.*, 1 (1), 1-6.

Rudneva, I. I., Skuratovskaya, E. N., Dorohova, I. I. and Kovyrshina, T. B. (2012). Use of fish blood biomarkersfor evaluating of marine environmenthealth. *World J. Sci. Technol.*, 2(7), 19-25.

Serafim, A., Company, R. B., Lopes, B., Fonseca, V. F. S., França, S. R. P., Vasconcelos, R. P., Bebianno, M. J., Cabral, H. N. (2012). Application of an integrated biomarker response index (IBR) to assess temporal variation of environmental quality in two Portuguese aquatic systems. *Ecological Indicators*, 19, 215-225.

Sergeyeva, N. G., Kolesnikova, E. A. and Mazlumyan, S. A. (2010). Taxonomic diversity of meiobenthos in the Sevastopol bays (the Crimean Black Sea coast). In:*Bidiversity and Sustainable Development: Proc. Internat. Sci. pract. Conf.*, May 19-22, 2010, Simferopol). (pp. 113-115). (in Russian).

Shcurova, E. S. and Malahova, L. V. (2012). Comparative evaluation of the chlorinated hydrocarbons pollution of the ecosystem of Laspy Bay and Marinov Bay (Black Sea) and the several regions of Mediterranean. In: M. E. Sokolova, G. A. Golubeva, V. A. Ivanova, etc. (Eds.), «Lomonosov Reading 2012.» 242 p. Sevastopol: «Express-print.» (*in Russian).*

Sole, M., Rodriguez, S., Papiol, V., Maynou, F., Cartes, J.E. (2009). Xenobiotic metabolism markers in marine fish with different trophic strategies and their relationship to ecological variables. *Comp. Biochem. Physiol.* 149C (1), 83-89.

Stalnaya,I.D., andGarishvily, T.G.(1977). Methodof malonic dyaldehydedetermination using thiobarbituric acid.In: *CurrentMethods inBiochemistry*. Moscow:Medicine Publ.66-68.(*in Russian).*

Svetovidov, A. N. (1964). *Black Sea Fishes*. Leningrad: Nauka. (in Russian).

The World Health Report (2002). World Health Organization, Geneva, Switzerland.

Van der Oost, R., Beyer, J. and Vermeulen N. P. E. (2003). Fish bioaccumulation and biomarkers in environmental risk assessment: a review. *Environ. Toxicol. Pharmacol.*, 13, 57-149.

Vladimirov, J. A. (2001). Activated chemiluminescence and bioluminescence as an instrument in medicine-biological studies. *Soros Educational Journal*, 7, 16-20. (*in Russian).*

In: Advances in Marine Biology. Volume 1 ISBN: 978-1-53610-131-7
Editors: Adam Kovács and Patrik Nagy © 2016 Nova Science Publishers, Inc.

Chapter 6

TROPHIC ECOLOGY OF THE SHORTFIN MAKO *ISURUS OXYRINCHUS* (LAMNIFORMES: LAMNIDAE) IN THE EASTERN PACIFIC OCEAN

Rigoberto Rosas-Luis**, Jonathan E. Pincay-Espinoza, Peggy Loor-Andrade and Maribel Carrera-Fernández

Departamento Central de Investigación, Universidad Laica "Eloy Alfaro" de Manabí, Ciudadela Universitaria vía San Mateo, Manta, Manabí, Ecuador

ABSTRACT

Shortfin mako sharks *Isurus oxyrinchus* are important organisms in the pelagic ecosystem. In spite of their vulnerable status, they are captured by fishermen in the Eastern Pacific Ocean. The proper management and conservation of this population in the Pacific is critical and thus, more biological knowledge of this species is needed. In this work, the trophic habits and consumption of key prey by *I. oxyrinchus* were studied through stomach content analysis of samples collected off the coast of Ecuador, and a literature review of its feeding habits was conducted for the eastern Pacific. The index of relative importance of prey items, and the consumption of key prey were calculated for the three areas. The diet was composed of 78 prey taxa, and cephalopods were the

* Corresponding author: riroluis@yahoo.com.mx.

main prey items in the north and central areas, with the jumbo squid *Dosidicus gigas* being the most important prey. For the southern area, *I. oxyrinchus* fed on fish with the bigeye cigarfish *Cubiceps pauciradiatus* being the most important prey species. The trophic niche of *I. oxyrinchus* was composed of squids and fish as the main prey items. The niche breadth for the three areas was reduced, and the trophic level calculated with stomach contents and stable isotopes of $\delta^{15}N$ showed values of 3.68 to 4.44 suggesting that this shark is a tertiary predator in the ecosystem. The jumbo squid *D. gigas* and fish of the family Scombridae were described as key species in the diet of *I. oxyrinchus*. The consumption estimate from 2000 to 2013 was 39 128 tons, with 55.8% of this consumption represented by *D. gigas*. These results demonstrate that *I. oxyrinchus* is an active and top predator in the trophic webs of the Eastern Pacific, with opportunistic feeding habits on the more available and abundant prey species.

Keywords: *Isurus oxyrinchus*, shortfin mako, trophic ecology, Eastern Pacific Ocean

1. INTRODUCTION

The shortfin mako shark *Isurus oxyrinchus* (Rafinesque, 1810) is a species of the family Lamnidae, which inhabits the waters of tropical and subtropical oceans (Compagno, 2001; Compagno et al. 2005; Nelson, 2006), with a distribution in the Atlantic, Indic and Pacific Oceans (Nelson, 2006). Sharks have a slow growth rate, late maturity, and long gestation period, biological characteristics that are sensitive to marine changes (Cortés, 2000; Frisk et al. 2001; Frisk et al. 2005).

In the early 1920s, the shark fisheries showed an increase in captures (Barker and Schluessel, 2005). As a result of this fishery activity, ocean shark populations have been declining (Dulvy et al. 2008; Ferreti et al. 2008; Mucientes et al. 2009; Dulvy et al. 2014). In spite of the fisheries regulation, the over-fishing continues as the main factor influencing the decline of sharks (Ferrati et al. 2010). According to the Food and Agriculture Organization of the United Nations (FAO, 2015), in the Eastern Pacific Ocean *I. oxyrinchus* is an important species in the total catches of sharks, and according to the International Union for the Conservation of the Nature (IUCN), it is considered a vulnerable species (Polidoro et al. 2012).

Sharks, such as *I. oxyrinchus*, are key top species that regulate the food webs in the ecosystems (Motta and Wilga, 2001; Navia et al. 2010;

Bornatowski et al. 2014). *Isurus oxyrinchus* is an active predator of bony fishes and cephalopods in the Atlantic Ocean and in the south Pacific off the coast of Chile (Stillwell and Kohler, 1982; Vaske-Júnior and Rincón-Filho, 1998; Cortés, 2000; Maia et al. 2006; Mucientes-Sandoval and Saborido-Rey, 2008; Lopez et al. 2009; Wood et al. 2009). Its diet may include other sharks, dolphins and marine mammals (Cliff et al. 1990; Velasco-Tarelo, 2005; Mucientes-Sandoval and Saborido-Rey, 2008; Lopez et al. 2009; Lopez et al. 2012; Preti et al. 2012; Biton-Porsmoguer et al. 2015). As *I. oxyrinchus* is a top predator, its depletion from the oceans promotes changes in top-down control (Ferreti et al. 2008; Ferreti et al. 2010), and modifies the structure and flux of biomass and energy in the ecosystems (Ferreti et al. 2008; Heupel et al. 2014). Thus, it is necessary to identify their food sources and feeding habits at a local and regional scale, and perform a global analysis to identify trends in the trophic ecology of this shark.

Research on the diet and feeding behavior, as well as the predation strategy, of sharks has been conducted in the North and South Pacific Ocean (Yatsu, 1995; Mucientes-Sandoval and Saborido-Rey, 2008; Vetter et al. 2008; Lopez et al. 2009). In the central Pacific off the coast of Ecuador, greater knowledge of *I. oxyrinchus* and other sharks is needed. Trophic ecology studies are conducted to determinate the function and position of species in the food web, and are important to understanding the impacts of predators on the abundance and biomass of trophic levels (Motta and Wilga, 2001; Aguiar and Valentin, 2010; Hussey et al. 2014). Thus, the objective of this research was to investigate biological aspects and the trophic ecology of *I. oxyrinchus* caught in the Ecuadorian Pacific and compare these results with those from the North and South Eastern Pacific Ocean, identifying key prey species, and inferring the biomass consumption of key prey species.

2. Methods and Statistical Analyses Applied in the Trophic Ecology of *Isurus oxyrinchus*

2.1. Stomach Content Analysis Applied to the Trophic Ecology of *Isurus oxyrinchus*

2.1.1. Prey Identification

The stomach contents of sharks have generally been collected from the by-catch of the artisanal or industrial fleets (Stillwell and Kohler, 1982; Vaske-

Júnior and Rincón-Filho, 1998; Cortés, 2000; Maia et al. 2006; Mucientes-Sandoval and Saborido-Rey, 2008; Lopez et al. 2009; Wood et al. 2009; Galván-Magaña et al. 2013). Fishing methods used by fishermen are long-line and surface and middle water gillnet. The stomachs are extracted and their fullness is estimated as a percentage of stomach capacity. The stomach content is placed in a labeled bag, then keep on ice for transfer to the Laboratory, and frozen directly for later analysis. Prey are separated and identified to the lowest taxonomic level using different keys of identification and weighted with a precision of 0.1 g. For the identification of complete bony fishes, squids and cartilaginous fishes, the keys of Fischer et al. (1995a, b), and Jered and Roper (2010) are used. Fish remains are identified using the skeleton according to Clothier (1950), and the reference collection of fish skeletons of each institution. Hard structures such as otoliths and squid beaks are identified according to Wolff (1984), North et al. (1984), Clarke (1986), García-Godos (2001) and Xavier and Cherel (2009). Additional, the total length or fork length with precision in centimeters and the sex of each shark are recorded.

2.1.2. Cumulative Prey Curves

Cumulative prey curves have been constructed to determine if an adequate number of stomachs had been analyzed. These curves are constructed at all taxonomic levels and excluded those prey that occurred only once (Preti et al. 2012). A vegan package (Oksanen et al. 2010) of the statistical software R (R Development Core Team, 2010) is used for this analysis. When the curve reached the asymptote, the sample size is considered sufficient to describe the diet (Hurtubia, 1973). When the asymptote is not evident a straight line to the last 4 points is fitted and compared the slope of the line with a line of slope zero, with the asymptote indicated when the lines did not differ significantly (Bizzarro et al. 2007).

2.1.3. Diet Analysis

To evaluate the feeding rate, vacuity index has been used (Consoli et al. 2008), with the following equation $V = N_e/N_s*100$, where, N_e = number of empty stomachs and N_s = total number of stomachs analyzed.

Stomach contents have been analyzed by frequency of occurrence, and numeric and gravimetric methods to quantify the diet. Frequency of Occurrence (%FO) is calculated as the percentage of predators that fed on a certain prey. The number of individuals of a certain prey relative to the total number of individual prey is represented by Number (%N), and the weight of a certain prey relative to the total weight of all prey is calculated as Weight

(%W) (Cailliet, 1976). The percentage of the Index of Relative Importance (%IRI) is calculated to describe the contribution of each prey in the diet (Pinkas et al. 1971), as follows %IRI = ((IRI*100)/∑IRI). Graphics of the most important prey items by area have generally been plotted.

Comparisons between groups have been tested using contingency tables and by estimating G statistics (Crow, 1981). This statistic has a $\chi 2$ distribution.

$$G = 2 * \sum_{i.j} X_{ij} \ln\left(X_{ij} / \left(X_i X_j / N\right)\right)$$

Where Xij is the FO of prey of the i category ingested by the j predator category, Xi is the FO of prey of the i category ingested by all predators, Xj is the total FO of prey ingested by the j predator category, and N is the total number of prey items ingested by all predators.

2.1.4. Trophic Overlap

The index of Morisita-Horn has been used to calculate the trophic overlap between the food and feeding habits, as follow, $C\lambda = 2\sum(Pxi*Pyi)/(\sum Pxi^2 + \sum Pyi^2)$ where $C\lambda$ is the Morisita-Horn index between the x and y predators, Pxi is the proportion of prey i in the total prey consumed by predator x, Pyi is the proportion of prey i in the total prey consumed by predator y, n is the number of total prey (Smith and Zaret, 1982). Values of this index vary from 0 to 1, where values >0.6 indicate high overlap, 0.3 to 0.6 indicate a medium overlap, and 0.0 to 0.29 indicate low overlap (Langton, 1982).

2.1.5. Niche Breadth and Trophic Level

The %IRI has been used to obtain the niche breadth (Bi) by calculating the standardized Levin's index (Hurlbert, 1978; Krebs, 1999), as follows $Bi = ((1/\sum Pij^2))-1)/(n-1)$, where Bi is the Levin's index of predator i, Pij is the diet proportion of predator i that uses prey j and n is the number of prey categories. This index takes the values of 0 to 1; when $Bi < 0.6$ the niche breadth is narrow, and when $Bi > 0.6$ the niche breadth is wide (Krebs, 1999).

The trophic level has been calculated with the following equation: $TL_k = 1 + (\sum Pj*TLj)$, where TL_k is the trophic level, Pj is the proportion of the prey category in the predator diet, TLj is the trophic level of the prey categories (bony fishes $TLj = 3.24$, cephalopods $TLj = 3.20$, crustaceans $TLj = 2.52$, invertebrates $TL\,j = 2.5$, seabirds $TLj = 3.87$, chondrichthyan fishes TLj=3.65 and marine mammals $TLj = 4.02$) (Cortés, 1999).

2.1.6. Feeding Strategy

The feeding strategy of *I. oxyrinchus* has been described using the method of Amundsen et al. (1996). This is a two-dimensional method based on the specific abundance of prey (%SAP) and the frequency of occurrence (%FO). Results of this method describe the prey importance as rare or dominant, the feeding strategy as a generalist or specialist predator, and the inter- and intra-specific components of the niche breadth. The values are calculated with the following equation: $Pi = (\sum Si/\sum Sti)*100$, where Pi is the specific abundance of prey i, Si is the weight in grams of prey i, and Sti is the total weight of all stomach contents of stomachs with prey i. It was not possible to apply this method to the other two areas because there were no data on individual feeding.

2.1.7. Consumption by Sharks of Key Prey Species

Estimates of consumption by sharks have been calculated using catches as a measure of abundance in the ocean because biomass estimates of sharks are not available. The analysis assumes that the proportion of each prey species does not vary over time. The consumption equation is represented as follows: $Qi = \sum Fj*(Q/B)*DCij$, where Q_i is the consumption of prey species, Fj represents fishery catches of sharks in tons, $(Q/B)j$ is the Consumption/Biomass relationship of sharks, and ($DCij$ = %IRI) is the diet proportion of each prey species i in the diet of sharks j.

2.2. Stable Isotopic Analysis

2.2.1. Sampling and Isotopic Analysis

A small portion of muscle (dorsal portion of the head, or caudal portion) is cleaned with distilled water and lipids are extracted with chlorophorm and methanol (Bligh and Dyer 1959). Samples are freeze-dried and powdered, and 0.285-0.315 mg of each sample is packed into tin capsules. Isotopic analyses are performed at different laboratories (Laboratory of Stable Isotopes at the Estación Biológica de Doñana, Plant Science Department at University of California, Laboratorio de Isótopos estables UNAM, Laboratorio de Isotópos Estables CICIMAR-IPN). Samples are generally heated to 1020°C using a continuous flow isotope-ratio mass spectrometer (Thermo Electron) by means of a Flash HT Plus elemental analyser interfaced with a Delta V Advantage mass spectrometer. Stable isotope ratios are expressed in the standard δ-notation (‰) relative to Vienna Pee Dee Belemnite ($\delta^{13}C$) and atmospheric N_2

($\delta^{15}N$). Based on laboratory standards, the measurement error is ± 0.1 and ± 0.2 for $\delta^{13}C$ and $\delta^{15}N$, respectively. The C:N ratio of tissues has to be always lower than 3‰. ANOVA tests have been used when comparing isotope values between groups.

2.2.2. Trophic Level based on Isotope Values

The trophic level position of each species has been calculated according to Kim et al. (2011) and Carlisle et al. (2012), using the following equations:

$$TL = [(\delta^{15}N_{predator} - \delta^{15}N_{base})/TEF] + 3,$$

where TL is the trophic level of that consumer, $\delta^{15}N_{predator}$ represents the values of the consumer´s muscle tissue; $\delta^{15}N_{base}$ is the mean $\delta^{15}N$ of *Benthosema panamense* (9.9‰; for the Pacific off Ecuador; Calle, 2010, with a trophic level of 3; Dambacher et al., 2010), and *Pleuroncodes planipes* (12.3‰; for the north Pacific with a trophic level of 2.6; Madigan et al., 2012) as reference species that share the same habitat as predators, and TEF = 3.7 is the transfer energy factor of each predator (Sweeting et al. 2007).

2.3. Trophic Ecology of *Isurus oxyrinchus* in the Eastern Pacific

2.3.1. Feeding Habits of Isurus oxyrinchus

For description of the feeding habits of *I. oxyrinchus* off Ecuadorian waters (central Pacific), samples were collected in the fishing port of Santa Rosa, Santa Elena, Ecuador 02°13´S; 80°58´W, from July 2013 to April 2015, (Figure 1). For the north and south Pacific, reports of the food and feeding habits of *I. oxyrinchus* were reviewed. For the northern area of Eastern Pacific Ocean, the reports of Vetter et al. (2008) and Preti et al. (2012) were consulted, and for the southern area, the reports of Yatsu, (1995), Mucientes-Sandoval and Saborido-Rey, (2008), Lopez et al. (2009), Lopez et al. (2012), and Rosas-Luis et al. (2016) were consulted. To perform an adequate comparison of the prey items of *I. oxyrinchus*, the diet values reported off the coast of United States of America by Lopez et al. (2009), and off the coast of Chile by Preti et al. (2012) and the present study (Figure 1) were used to compare and estimate the feeding habits and consumption of sharks.

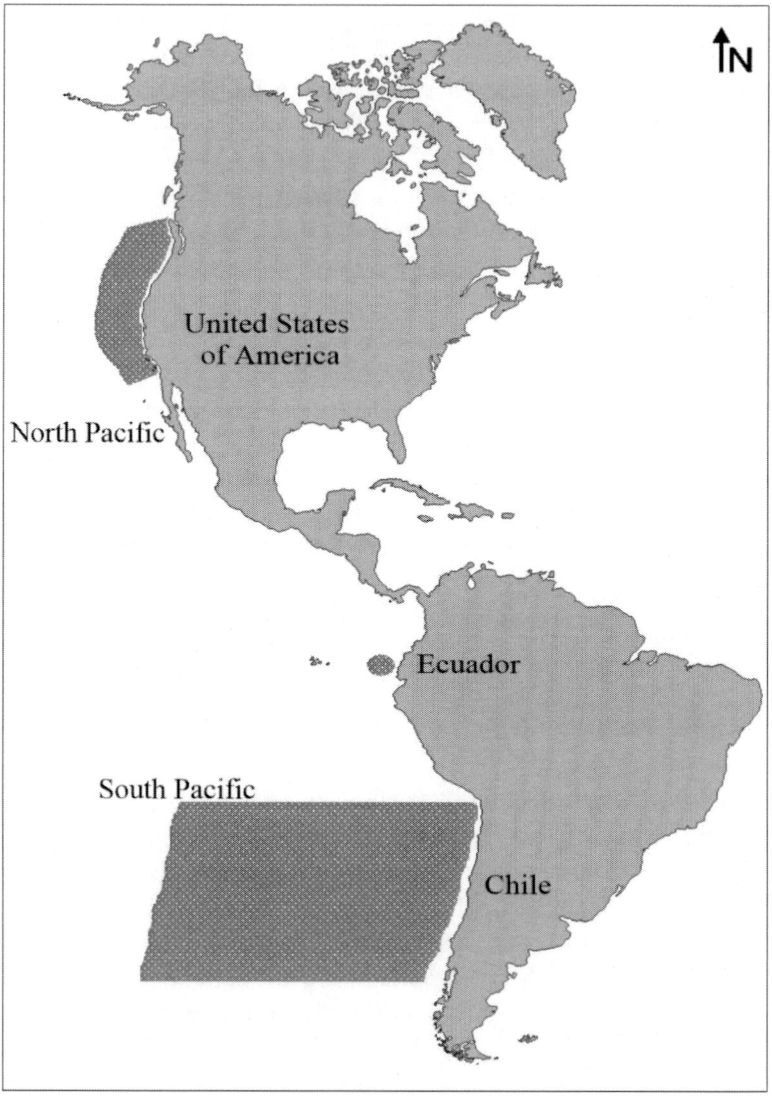

Figure 1. Eastern Pacific Ocean. Gray dark polygons represent the northern (Off the coast of the United States of America), central (off the coast of Ecuador), and southern (off the coast of Chile) areas where stomach content analyses were reported for *Isurus oxyrinchus*.

The consumption by *I. oxyrinchus* of the jumbo squid *D. gigas* and fishes of the family Scombridae (all scombrid fishes were unified into a single group; species included were *K. pelamis*, *T. albacares*, *Auxis* spp., *Scomber japonicus*, *Sarda chiliensis*, and unidentified scombrid fishes) were calculated for the period 2000 to 2013 based on the available fishery data of each area. The Q/B value was taken from Rosas-Luis et al. (2008).

2.3.2. Stable Isotopic Analysis of Isurus Oxyrinchus

Stable isotopic analysis of $\delta^{15}N$ and $\delta^{13}C$ was applied to muscular samples collected in the fishing port of Santa Rosa, Santa Elena, Ecuador 02°13´S; 80°58´W, from July 2013 to April 2015. Reported data of Malpica-Cruz et al. (2013) and Maya-Meneses et al. (2016) for the North Pacific, and Clarian et al. (2016) for the south Pacific, were consulted and compared with those of the central area. Three size groups were used in the comparison, small sharks <112 cm total length, medium sharks 112-196 cm, and large sharks >196 cm.

3. Feeding Description of *Isurus Oxyrinchus* in the Eastern Pacific Ocean

3.1. Results of Stomach Content Analysis

3.1.1. Central Area "Ecuadorian Pacific"

General Description of *I. oxyrinchus* Caught in the Central Area

The feeding habits of 281 shortfin mako sharks were analyzed from the Ecuadorian Pacific Ocean: 134 were males, 129 females and 18 undetermined, with a sex proportion of 1.04M:1H ($\chi^2 = 0.095$; $p = 0.75$). The size of males ranged from 106 to 257 cm total length (Mean = 174.51 ± 23.60) and females from 105 to 267 cm total length (Mean = 176.16 ± 25.40) (Figure 2). The undetermined individuals ranged from 108 to 213 cm total length, and no size differences were observed among males and females ($t = -0.54$; $p = 0.58$). Figure 3 shows the cumulative prey curve. When all the prey items were included, the curve did not reach the asymptote ($R^2 = 0.95$, $p = 0.01$), but did when prey that occurred once were excluded ($R^2 = 0.85$, $p = 0.05$).

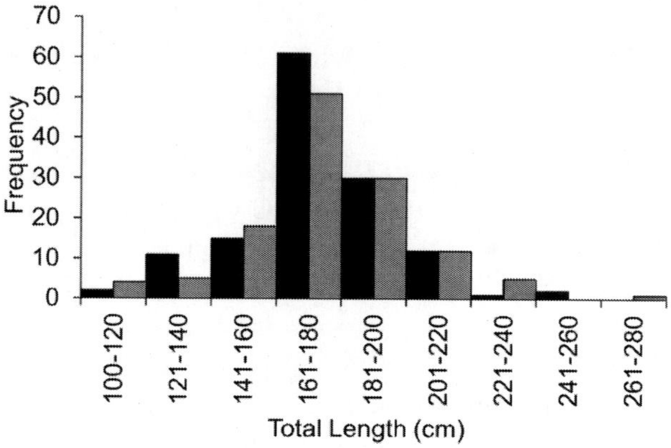

Figure 2. Size distribution of the shortfin mako *Isurus oxyrinchus* in centimeters (cm) caught off the coast of Ecuador. Black bars represent males and dark gray bars represent females.

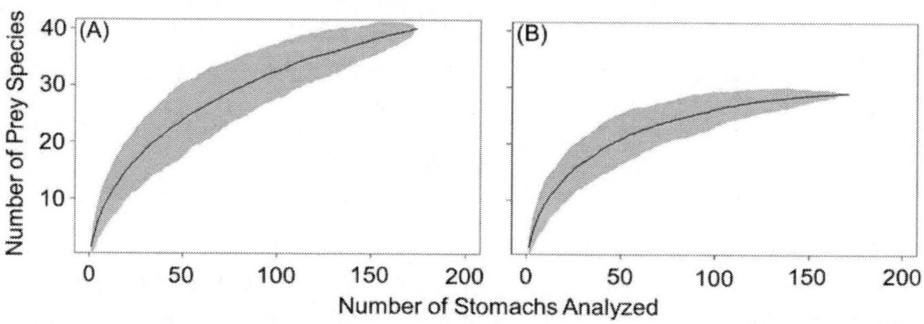

Figure 3. Cumulative curve of prey (black line) and standard deviation (gray) calculated for *Isurus oxyrinchus* caught off the coast of Ecuador. (A) represents the curve when all stomachs analyzed with all prey items were included, and (B) represents the curve calculated when uncommon prey were deleted (prey that occurred once in the stomachs).

Prey Identification

Forty taxa were identified in the stomach contents of *I. oxyrinchus* caught in the Ecuadorian Pacific (Table 1). Cephalopoda was the main prey group (%IRI = 58.28%), with squids of the family Ommastrephidae being the most important in the diet (%IRI = 48.36). The second most important group in the diet was Osteichthyes (%IRI = 41.07%), with fishes of the family Scombridae being the most important (%IRI = 20.62) (Figure 4, Table 1). The jumbo squid *D. gigas* was the main prey species in the diet (%IRI = 48.28), followed by the skipjack *K. pelamis* (%IRI = 11.39), the scombrid *Auxis* spp. (%IRI = 7.47), and finally, the squid *Ancistrocheirus lesueurii* (%IRI = 7.26) (Figure 4 and Table 1).

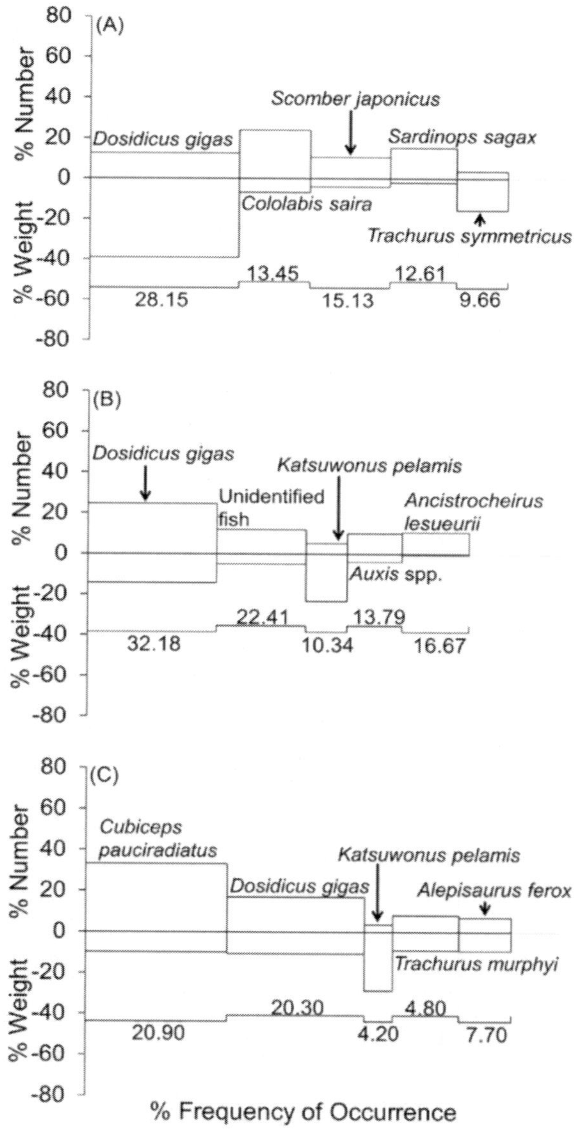

Figure 4. Diet composition of the main prey found in the stomach contents of *Isurus oxyrinchus* in the Eastern Pacific Ocean, North (A), Central (B) and South (C), expressed in the percentage of Number (%N) and Weight (%W) (vertical axis) and the Frequency of Occurrence (%FO) (horizontal axis).

Table 1. Index of Relative Importance (%IRI), Frequency of Occurrence (%FO), Number (%N) and Weight (%W) of prey in the stomach contents of shortfin mako *Isurus oxyrinchus* in the Eastern Pacific Ocean. Un = Unidentified. Northern and southern areas values were calculated based on Preti et al. (2012) and Lopez et al. (2009)

	Southern area				Central area				Northern area			
Prey species	%N	%P	%FO	%IRI	%N	%P	%FO	%IRI	%N	%P	%FO	%IRI
Crustacea												
Solenoceridae												
Haliporoides diomedae	0.60	1.00	0.70	0.05								
Cephalopoda												
Ancistrocheiridae												
Ancistrocheirus lesueurii					10.59	0.69	16.67	7.26				
Argonautidae												
Argonauta spp.									1.55	0.00	0.84	0.05
Bolitaenidae												
Japetella sp.									0.24	0.00	0.84	0.01
Enoploteuthidae												
Abraliopsis sp.									0.12	0.00	0.42	0.00
Gonatidae												
Gonatopsis borealis									0.48	0.00	1.26	0.02
Gonatus sp.					0.52	0.00	1.15	0.02	4.04	0.00	9.24	1.29
Histioteuthidae												
Histioteuthis dofleini					1.81	0.01	4.02	0.28	2.97	0.03	7.98	0.83
Histioteuthis heteropsis									0.48	0.00	1.68	0.03
Histioteuthis sp.	0.60	0.40	0.70	0.03	1.81	0.37	2.87	0.24	1.43	0.02	3.78	0.19
Loliginidae												

	Southern area				Central area				Northern area			
Prey species	%N	%P	%FO	%IRI	%N	%P	%FO	%IRI	%N	%P	%FO	%IRI
Loligo opalescens									1.31	0.00	3.36	0.15
Mastigoteuthidae												
Mastigoteuthis dentata					0.52	0.00	1.15	0.02				
Octopoteuthidae												
Octopoteuthis sicula					1.03	0.00	2.30	0.09				
Octopoteuthis sp.									1.55	0.00	3.78	0.20
Ommastrephidae												
Dosidicus gigas	16.90	10.70	20.30	26.19	24.55	14.3	32.18	48.28	12.6	38.97	28.15	50.21
Hyaloteuthis pelagica					0.78	0.05	0.57	0.02				
Ommastrephidae sp. 1					0.26	0.00	0.57	0.01				
Ommastrephidae sp. 2					0.26	0.00	0.57	0.01				
Sthenoteuthis oualaniensis					0.52	0.56	1.15	0.05				
Toradores filipovae	2.30	0.50	2.80	0.37								
Onychoteuthidae												
Onykia robusta									0.12	0.00	0.42	0.00
Onychoteuthis sp.					3.36	0.10	6.90	0.58	0.36	0.00	0.84	0.00
Opistoteuthidae												
Opistoteuthis sp.					0.26	0.02	0.57	0.01				
Thysanoteuthidae												
Thysanoteuthis rhombus					2.58	0.41	4.60	0.53				
Unidenfied Cephalopoda	5.80	2.70	7.00	2.78	2.84	0.83	6.32	0.89	2.62	0.50	5.88	0.56
Osteichthyes												
Alepisauridae												
Alepisaurus ferox	6.90	9.20	7.70	5.80								

Table 1. (Continued)

	Southern area				Central area				Northern area			
Prey species	%N	%P	%FO	%IRI	%N	%P	%FO	%IRI	%N	%P	%FO	%IRI
Belonidae												
Belonidae Un					4.13	1.66	1.72	0.39				
Bramidae												
Brama australis	0.60	0.10	0.70	0.02								
Brama japonica									0.95	4.77	2.94	0.58
Carangidae												
Caranx sp.					0.26	0.13	0.57	0.01				
Decapterus macarellus					0.26	0.03	0.57	0.01				
Decapterus macrosoma					0.52	0.09	1.15	0.03				
Decapterus sp.					0.52	0.29	1.15	0.04				
Trachurus murphyi	8.10	9.00	9.80	7.83								
Trachurus symmetricus									3.45	15.64	9.66	6.38
Clupeidae												
Sardinops sagax									15.2	2.11	12.61	7.56
Coryphaenidae												
Coryphaena hippurus					2.07	7.40	4.60	1.68				
Echeneidae												
Remora remora					0.26	0.40	0.57	0.01				
Engaulidae												
Engraulis mordax									1.31	0.00	2.10	0.10
Haemulidae												

	Southern area				Central area				Northern area			
Prey species	%N	%P	%FO	%IRI	%N	%P	%FO	%IRI	%N	%P	%FO	%IRI
Haemulidae Un					0.52	0.23	1.15	0.03				
Kyphosidae												
Medialuna californiensis									0.12	0.00	0.42	0.00
Macrouridae												
Coryphaenoides acrolepis									0.24	0.38	0.84	0.02
Coryphaenoides sp.									0.24	0.00	0.84	0.01
Merlucciidae												
Merluccius gayi					0.26	0.03	0.57	0.01				
Monacanthidae												
Aluterus monoceros					0.26	0.41	0.57	0.01				
Mugilidae												
Mugil cephalus					0.26	0.18	0.57	0.01	3.33	12.46	10.5	5.73
Nomeidae												
Cubiceps pauciradiatus	33.10	9.80	20.90	41.91								
Cubiceps paradoxus					0.78	0.09	0.57	0.02				
Cubiceps sp.	1.70	1.10	2.10	0.27								
Ophidiidae												
Chilara taylori									0.12	0.00	0.42	0.00
Paralepididae												
Magnisudis atlantica									2.26	0.23	1.68	0.14
Paralepididae Un									0.59	0.14	1.26	0.03
Paralichthyidae												
Citharichthys sordidus									0.12	0.00	0.42	0.00

Table 1. (Continued)

	Southern area				Central area				Northern area			
Prey species	%N	%P	%FO	%IRI	%N	%P	%FO	%IRI	%N	%P	%FO	%IRI
Priacanthidae												
Pristigenys serrula					0.52	0.87	1.15	0.06				
Regalecidae												
Ragalecus glesne					0.26	1.33	0.57	0.04				
Scomberesocidae												
Cololabis saira									23.9	6.89	13.45	14.32
Scombridae												
Auxis spp.	0.60	0.90	0.70	0.05	10.08	3.97	13.79	7.47	0.24	0.98	0.42	0.02
Katsuwonus pelamis	3.50	28.8	4.20	6.34	5.17	23.4	10.34	11.39	0.36	0.99	1.26	0.06
Sarda chiliensis									0.48	0.97	1.26	0.06
Scomber japonicus	6.40	2.60	7.70	3.24	0.52	0.26	1.15	0.03	10.6	4.09	15.13	7.68
Scombridae Un	0.60	1.00	0.70	0.05	0.78	1.97	1.72	0.18	0.12	0.08	0.42	0,.03
Thunnus albacares					1.55	12.3	2.87	1.54				
Scorpaenidae												
Sebastes jordani									0.71	0.79	0.42	0.02
Sebastes goodei									0.12	0.11	0.42	0.00
Sebastes spp.									0.24	0.04	0.84	0.01
Tetraodontidae												
Lagocephalus lagocephalus					5.94	4.06	9.20	3.55				
Xiphiidae												

	Southern area				Central area				Northern area			
Prey species	%N	%P	%FO	%IRI	%N	%P	%FO	%IRI	%N	%P	%FO	%IRI
Xiphias gladius	1.20	1.50	1.40	0.18								
Unidenfied Osteichthyes	6.40	5.30	7.00	3.83	11.89	4.96	22.41	14.57	4.88	3.31	12.61	3.57
Chondrichthyes												
Carcharhinidae												
Prionace glauca									0.24	3.48	0.84	0.11
Lamnidae												
Isurus oxyrinchus	1.20	2.10	1.40	0.22								
Lamna nasus	1.20	7.40	1.40	0.56								
Mobulidae												
Mobula japanicus					0.26	13.8	0.57	0.31				
Pseudocarchariidae												
Pseudocarcharhias kamoharai	0.60	2.20	0.70	0.09								
Unidentified Chondrichthyes	0.60	0.80	0.70	0.05								
Delfinidae												
Delphinus delphis									0.12	2.88	0.42	0.04
Delfinidae Un					0.52	2.25	1.15	0.12				
Tursiops truncatus	0.60	3.50	0.70	0.13								
Unidentified mammals					0.78	2.45	1.72	0.21				

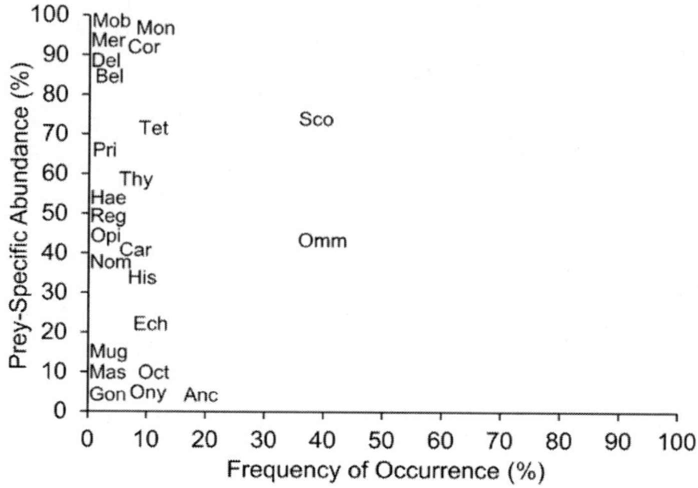

Figure 5. Diagram of the feeding strategy of the shortfin mako *Isurus oxyrinchus* in the Pacific Ecuadorian, expressed in the Specific Prey Abundance and the Frequency of Occurrence of family prey. Bel: Belonidae, Car: Carangidae, Cor: Coryphaenidae, Ech: Echeneidae, Hae: Haemulidae, Mer: Merluciidae, Mon: Monacanthidae, Mug: Mugilidae, Nom: Nomeidae, Pri: Priacanthidae, Reg: Ragalecidae, Sco: Scombridae, Tet: Tetraodontidae. Anc: Ancistrocheiridae, Gon: Gonatidae, His: Histioteuthidae, Mas: Mastigoteuthidae, Oct: Octopoteuthidae, Omm: Ommastrephidae, Ony: Onychoteuthidae, Opi: Opistoteuthidae, Thy: Thysanoteuthidae. Mob: Mobulidae. Del: Delfinidae.

3.1.2. Northern and Southern Pacific Areas

Prey Identification

Values of %IRI, %FO, %N from the north and south Pacific areas are shown in Table 1. For the northern area, 42 taxa have been reported in the diet of *I. oxyrinchus*. Cephalopods have been reported as the main prey items, with squid *D. gigas* of the family Ommastrephidae being the most important, followed by the group composed of Osteichthyes, with the fish *Cololabis saira* of the family Scomeresocidae and the chub mackerel *Scomber japonicus* of the family Scombridae being the most important. Other prey groups such as Chondrichthyes, mammals, and marine birds have also been reported (Figure 4, Table 1).

Twenty-two prey taxa have been reported in the diet of *I. oxyrinchus* in the southern area (Table 1). Osteichthyes have been reported as the main prey group, with the fish *Cubiceps puciradiatus* of the family Nomeidae being the most important. The second most important group was Cephalopoda, with the jumbo squid *D. gigas* of the family Ommastrephidae being the most important. Other groups of less importance such as Chondrichthyes, Mammalia and Crustacea have also been reported in the diet (Table 1, Figure 4).

3.1.3. Trophic Overlap

Prey items, found in the stomach contents of sharks from the Ecuadorian Pacific and those reported for the northern and southern Pacific, overlap with five specific prey (*Katsuwonus pelamis, Auxis* sp., *S. japonicus*, *D. gigas*, *Histioteuthis* sp.). Based on these prey items, a high trophic overlap was found between the northern and central areas $C\lambda = 0.87$, while between the central and southern areas, and the northern and southern areas, trophic overlap was moderate ($C\lambda = 0.52$, and $C\lambda = 0.48$ respectively).

3.1.4. Niche Breadth and Trophic Level

The niche breadth for *I. oxyrinchus* in the north was $Bi = 0.06$ with 42 prey taxa. In the central area, it was $Bi = 0.07$ with 40 prey taxa, and for the south it was $Bi = 0.14$ with 21 prey taxa. The trophic level calculated for *I. oxyrinchus* in the northern area was 4.22, and for the central and southern areas, the trophic level was 4.23.

3.1.5. Feeding Strategy

The Amundsen graphical analysis, considering 24 prey families, indicated that most of the families found in the stomach contents, such as Echeneidae, Mugilidae, Gonatidae, Mastigoteuthidae, Octopoteuthidae and Onychoteuthidae, were consumed as rare items. The families Ommastrephidae and Scombridae showed the greatest abundance and frequency of occurrence in the Ecuadorian Pacific Ocean.

3.1.6. Consumption of Fish and Squid by Isurus oxyrinchus

Figure 6 shows the consumption by *I. oxyrinchus* of specific prey. The total consumption of fish in the three areas during the period analyzed was 39 128 tons. For squid, consumption was calculated to be 21 839 tons. The consumption of scombrid fishes was 6 657 total tons, and this value represented 17.01% of the total fish consumption. The total consumption of the jumbo squid *D. gigas* was 19 299 representing 88.37% of the total consumption of squids.

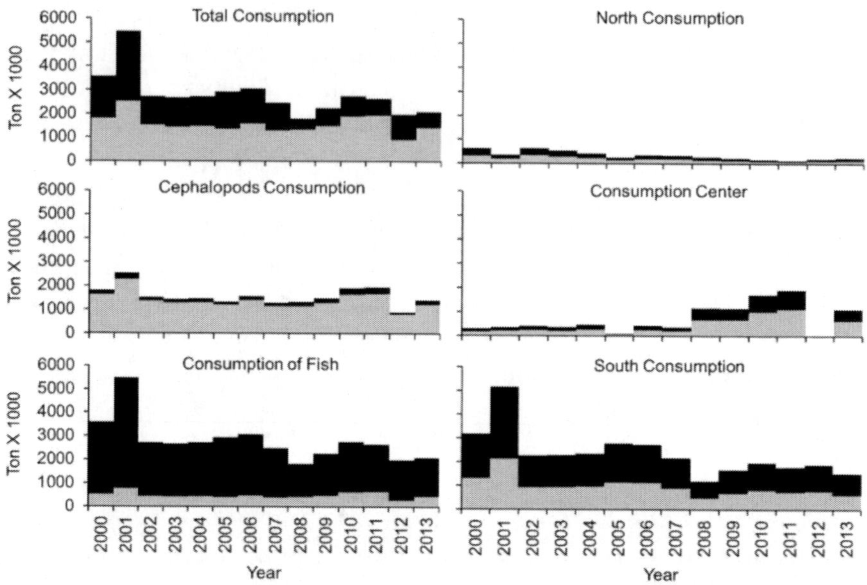

Figure 6. Total consumption of *Isurus oxyrinchus* in the Eastern Pacific Ocean: black area represents the consumption of fish, and gray area represents the consumption of cephalopods. Cephalopod Consumption in the Eastern Pacific Ocean: black area represents the total consumption, and gray area is the consumption of *Dosidicus gigas*. Consumption of fish in the Eastern Pacific Ocean: black area represents the total consumption, and gray area represents the consumption of Scombrid fish. Consumption of fish and cephalopods in the Eastern Pacific, North (United States of America), Central (Ecuador) and South (Chile): black area represents fish, and gray area represents cephalopods.

3.2. Results of Stable Isotope Analysis

$\delta^{15}N$ values of *I. oxyrinchus* of the central area (own unpublished data) were higher for sharks of medium size than larger size (one-way ANOVA, $F_{19} = 4.67$, $P < 0.05$)(Figure 7). Results of the comparison between N values of all size groups in the three areas showed no difference (one-way ANOVA, $F_6 = 2.46$, $P > 0.05$).

Trophic position of *I. oxyrinchus* based on $\delta^{15}N$ isotopes showed the highest values for medium sharks, and the lowest values for small sharks in the central and north areas (Table 2). The highest value of the trophic position was calculated for the south area, but no size class was reported. The trophic position values varied around 1 unit when the prey at the base of the food web changed (Table 2).

Values of $\delta^{13}C$ between medium and large sharks for the central area (own unpublished data) did not vary (one-way ANOVA, $F_{19} = 1.78$, $P > 0.05$). Results of the comparison of $\delta^{13}C$ values between the three areas did not vary (one-way ANOVA, $F_6 = 0.54$, $P > 0.05$) (Figure 7).

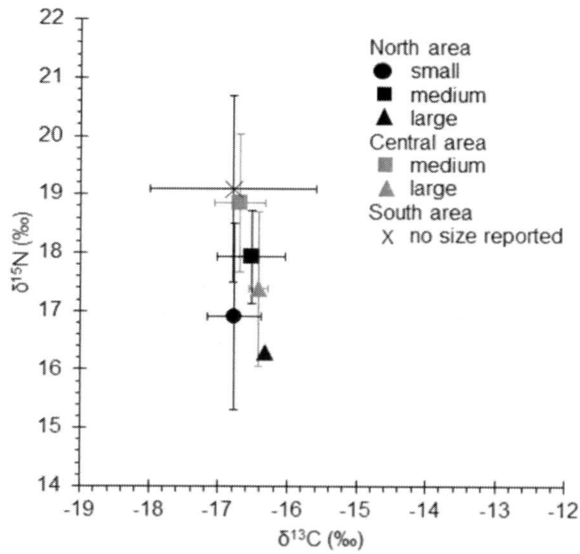

Figure 7. $\delta^{15}N$ and $\delta^{13}C$ values (mean ± s.d.) for *Isurus oxyrinchus* of the eastern Pacific ocean.

Table 2. Trophic position (TP) of *Isurus oxyrinchus* by size interval and studied areas. TP-1 was calculated by using the pelagic red crab *Pleuroncodes planipes* as base of the trophic web, and TP-2 was calculated by using the myctophid *Benthosema panamense*. * no size reported

Pacific area	Size	Number of samples	TP-1	TP-2
North	small	15	3.84	4.89
	medium	4	4.12	5.17
	large	6	3.68	4.73
Central	medium	41	4.37	5.42
	large	1	3.97	5.02
South	*	50	4.44	5.49

4. Discussion

The high metabolic rates of Lamnid sharks (Newton et al. 2015) make the identification of food items difficult. However, this can be mitigated if sharks are feeding when they are caught. The stomachs analyzed in each study showed that the sharks fed actively before they were caught, and it was confirmed with the low vacuity percentage of stomachs for the three areas of the Eastern Pacific Ocean. Moreover, the feeding behavior of *Isurus oxyrinchus* is characterized by the consumption of complete or large pieces of prey (Wood et al. 2009), further improving the identification of stomach contents. In this case, prey species were well-preserved and in earlier digestion stages, improving the identification of ingested species. The number of stomachs was adequate to describe the diet in the three areas: the United States of America, the South Pacific off the coast of Chile, and the Central Pacific Ocean off the coast of Ecuador.

4.1. Feeding Habits in the Eastern Pacific Ocean

The diet of *I. oxyrinchus* from different regions was composed of a similar number of prey items. Forty prey items were identified in the stomachs of sharks from the central area and 42 were reported for sharks off the coast of the United States (Preti et al. 2012). These may indicates the availability and abundance of similar prey in these areas, such as squids and fish (Clarke, 1996; Ebert and Stehmann, 2013; Galván-Magaña et al. 2013).

The feeding habits of *I. oxyrinchus* showed that these sharks are predators of fishes and squids, in agreement with the results previously reported for this species in the Pacific Ocean (Stillwell and Kohler, 1982; Velasco-Tarelo, 2005; Mucientes-Sandoval and Saborido-Rey, 2008; Lopez et al. 2009, 2012; Preti et al. 2012; Galván-Magaña et al. 2013). High consumption of bony fishes by *I. oxyrinchus* in the Ecuadorian Pacific and similar feeding behavior in the southern area suggests that these prey are abundant in the ecosystems and available to sharks (Stillwell and Kohler 1982; Lopez et al. 2009, 2012; Gorni et al. 2012, 2013). Nevertheless, as indicated by our results, the FO of squid was similar to fish, suggesting that these cephalopods are also important in their diets (Preti et al. 2012). The importance of fish and squid appears to vary between sampled areas and among seasons. There are reports showing fishes as the most important food source, indicating a feeding strategy based on ichthyophagous habits (Stillwell and Kohler, 1982; Velasco-Tarelo, 2005;

Mucientes-Sandoval and Saborido-Rey, 2008, Lopez et al. 2009, 2012; Preti et al. 2012). In other areas, teutophagous habits are suggested (Galván-Magaña et al. 2013).

There is evidence of a trophic overlap between the food sources used by *I. oxyrinchus* since the squid *D. gigas* and the fishes *K. pelamis* and *Auxis* spp. were the most important prey in the Ecuadorian Pacific and in the northern area off the US coast (Velasco-Tarelo, 2005; Mucientes-Sandoval and Saborido-Rey, 2008; Vetter et al. 2008; Lopez et al. 2009; Preti et al. 2012). Squids are consumed by other sharks such as *Alopias pelagicus*, *Alopias superciliosus*, and *Prionace glauca* in Ecuadorian waters, and three species have been reported in their diet: *D. gigas* and *Sthenoteuthis oualaniensis* of the family Ommastrephidae, and *Ancistrocheirus lesueurii* of the family Ancistrocheiridae (Galván-Magaña et al. 2013; Rosas-Luis et al. 2016). The use of *D. gigas* as prey by *I. oxyrinchus* may be a result of the sharks´ vertical migration to deep waters during the day, which can be up to 800 m deep (Bress, 1993; Abascal et al. 2001; Loefer et al. 2005; Field et al. 2008). These deep waters are where *D. gigas* is found (Nigmatullin et al. 2001; Bazzino et al. 2010). On the other hand, the use of scombrid fishes such as *K. pelamis* and *Auxis* spp. may be a result of their high abundance and distribution in the Pacific (Allen and Robertson, 1998). Contrary to the results for the northern and central areas where cephalopods are the most important prey, in the southern area the bigeye cigarfish *Cubiceps pauciradiatus* was reported as the main prey source (Lopez et al. 2009). However, as presented in this work, the second most important prey was *D. gigas*. The inclusion of cephalopods in the diet of *I. oxyrinchus* in the southern area varies year to year, as they were reported as the main prey item in 2008 (Mucientes-Sandoval and Saborido Rey, 2008), and were not important in 1995 (Yatsu, 1995). The important prey species in the diet suggests that *I. oxyrinchus* is an opportunistic feeder, and that its diet may change as a consequence of movements and abundance of prey.

4.2. Variation in Diet in the three Areas of the Eastern Pacific Ocean

Crustaceans, marine birds, rays, sharks, and marine mammals are uncommon prey species in the diet of *I. oxyrinchus*. Crustaceans and birds have been reported as prey sources of this shark for the southern and northern areas (Vetter et al. 2008; Lopez et al. 2009; Preti et al. 2012), but they were not found in the stomachs analyzed for the central area. It is known that crustaceans, such as the pelagic red crab *Pleuroncodes planipes* in the north Pacific and *Pleuroncodes monodon* in the southern hemisphere, are abundant (Aurioles-Gamboa 1992; Roa et al. 1995; Robinson et al. 2004) and preyed upon by large predators such as *D. gigas* (Rosas-Luis et al. 2008). Thus, these prey are indirectly consumed by *I. oxyrinchus* via squid predation. Birds as food sources are preyed upon in the southern hemisphere by sharks such as *Galeocerdo cuvier* (Cliff and Dudley, 1991; Heithaus, 2005), but for *I. oxyrinchus* these prey sources may be ingested when they are dead and floating on the surface waters.

The presence of uncommon prey species in the diet of *I. oxyrinichus* in each area suggests that the primary distribution of these prey species overlaps with that of *I. oxyrinchus*. A large manta ray *M. japonica* was found in the stomach contents of a shark, and some large sharks had consumed small marine mammals. However, as our results and those previously reported by Lopez et al. (2009) and Preti et al. (2012) showed, these prey were consumed primary by large sharks. The consumption of these prey types by *Carcharodon carcharias*, *Carcharhinus leucas*, *Galeocerdo cuvier*, *Hexanchus sixgill* and *Notorynchus cepedianus* suggests that these species are active predators of marine mammals and birds (Lowe et al. 1996; Heithaus, 2002; Moteiro et al. 2006; Bornatowski et al. 2012). Considering that marine mammals and large fish were consumed in low proportions by *I. oxyrinchus*, we cannot conclude that this shark is an active predator of this kind of prey in the eastern Pacific Ocean. Finally, the consumption of other sharks by *I. oxyrinchus* was found in the three studied areas in agreement with similar reports in the Atlantic and Indian Oceans (Cliff et al. 1990; Maia et al. 2006).

4.3. Niche Breadth and Trophic Level

Measures of trophic position and niche breadth are essential in the interpretation of trophic ecology (Hussey et al. 2014). These results demonstrated that *I. oxyrinchus* in the Eastern Pacific Ocean has a narrow niche breadth, as a result of a selectivity for prey of the families Scombridae and Ommastrephidae. Its estimated trophic level (4.23) reflects its high level of predation activity in the ecosystem. This is a value similar to other Lamnidae species such as *Carcharodon carcharhias* and *Lamna nasus* with high trophic levels of 4.5 and 4.2, respectively (Cortés, 1999), that indicates, *I.*

oxyrinchus is a top predator and a tertiary consumer in the ecosystem (Cortés, 1999; Estrada et al. 2003).

4.4. Feeding Strategy

The feeding strategy of *I. oxyrinchus* has been conditioned by prey availability, as this shark is a highly migratory species that exploits different areas and consumes a wide range of prey (Lopez et al. 2009). In the north (Velasco-Tarelo, 2005; Preti et al. 2012) and central Pacific, *I. oxyrinchus* fed mainly on *D. gigas, Katsuwonus pelamis*, *Auxis* spp., *Thunnus albacares* and *Scomber japonicus*, coinciding with the high abundance and distribution of these species in the Pacific Ocean (Punsly and Deriso, 1991; Markaida and Gilly, 2016). Nevertheless, feeding habits differed in the southern area off the coast of Chile where this consumer showed a heterogeneous and generalized diet (Lopez et al. 2009, 2012). This suggests that the niche breadth may vary according to geographic area, and that this shark preys opportunistically on abundant and available prey species.

4.5. Consumption of Fish and Squids by *Isurus oxyrinchus*

Isurus oxyrinchus is a top predator in the eastern Pacific Ocean, with an active vertical and horizontal migration (Fowler, 2014) that determines its impact on the food web. Captures of this shark for the analyzed areas are the only indicators permitting calculations of the trophic consumption of prey (Rosas-Luis et al. 2016). Considering this indicator, the consumption of fish is higher than squid, but as our results showed, these values vary between northern, central and southern areas. We focused our study on the most important prey to demonstrate how the decline of this shark may affect the abundance of other marine organisms (Chang and Liu 2009; Ferreti et al. 2010). If the actual tendency in the decline of *I. oxyrinchus* continues, the population of *D. gigas* and other prey will be affected positively. On the contrary, if the prey populations increase, the population of sharks may be affected positively. It is known that the estimates of biomass between prey and predator are overestimates (Jennings and Collingridge, 2015), but these can be used to create models and measurements for the management and conservation of endangered species.

The high consumption rates of the jumbo squid *D. gigas* by *I. oxyrinchus* suggests that this squid is an important component in the food web (Nigmatullin et al. 2001; Rosas-Luis et al. 2008; Alegre et al. 2014), and confirms its abundance and distribution patterns in the Eastern Pacific Ocean (Markaida and Gilly, 2016). On the other hand, the importance of scombrid fishes in the diet increases latitudinally, confirming their wide distribution in this region (Allen and Robertson, 1998), and its availability as a food source for *I. oxyrinchus*. So that, the trophic relationships between prey and predator are fragile, and small disturbances may affect both components in the ecosystem, resulting in disturbances throughout the trophic webs.

4.6. Stable Isotopic Analysis of *Isurus oxyrinchus*

Values of $\delta^{15}N$ and $\delta^{13}C$ of *I. oxyrinchus* in the central, north, and south eastern Pacific were interpreted in addition of the results of the analysis of stomach contents. Isotopes resulted in no difference between sharks sampled in the three areas. This result suggests that *I. oxyrinchus* has similar feeding habits in the whole area, being cephalopods and fish the main prey groups (Stillwell and Kohler, 1982; Velasco-Tarelo, 2005; Mucientes-Sandoval and Saborido-Rey, 2008; Lopez et al. 2009, 2012; Preti et al. 2012; Galván-Magaña et al. 2013). Malpica-Cruz et al. (2013) and Maya-Meneses et al. (2016) mentioned that small and medium size *I. oxyrinchus*, caught off the coast of Mexico, had no variation in the $\delta^{15}N$ values, contrasting with the result of the central area where medium and large *I. oxyrinchus* had significant variation. The difference in $\delta^{15}N$ values for sharks in the central area results of the ontogenetic shift to a broader diet (Preti et al. 2012), and a more diverse feeding preferences of medium size sharks (Malpica-Cruz et al. 2013). The trophic position of *I. oxyrinchus*, based in $\delta^{15}N$ values of *P. planipes* as the base of the food web, coincided with the trophic position calculated with the results of stomach content analysis and that reported by Cortés, (1999), but it was different when the myctophid *B. panamense* was used as the base of the food web. It suggest that the selection of the organism as base of the food web is determinant for the trophic position in the ecosystem, and that more research is needed in order to define similar organisms as key species for isotopic analysis in each studied area. Despite this, the trophic position of *I. oxyrinchus* vary according to size, habitat and availability of food, and caution must be consider when make conclusions of the trophic position of this species.

Regarding the $\delta^{13}C$ values, no differences were found in comparison between size and areas, which confirms a same use of habitat of this species in the eastern Pacific. *Isurus oxyrinchus* are known to migrate between inshore and offshore environments (Stillwell and Kohler, 1982), this patters have been documented when sex segregation is present (Maya-Meneses et al. 2016). Unfortunately, our results and those reported for the south and north areas are not adequate to perform an analysis by sex, and nothing can be concluded. Thus, isotopic concentrations of muscular tissue of females and males are needed in order to prove the sexual segregation in and offshore of this species.

CONCLUSION

In conclusion, our results demonstrate that the short fin mako shark *I. oxyrinchus* feeds mainly on bony fishes and squids, but its trophic habits include other prey such as rays, sharks, marine mammals, marine birds and crustaceans. As described, *I. oxyrinchus* feeds on abundant and available prey sources, demonstrating its role as an opportunistic feeder in the studied areas. Variation in the prey items identified in the stomach contents is a result of the latitudinal distribution of sharks and the distribution areas of the potential prey, but as these results show, the jumbo squid *D. gigas* and fish of the family Scombridae are the most important food sources across the region. The squid *D. gigas* may be considered a key species in the development of *I. oxyrinchus* populations as this squid has shown demographic explosions and traverses both hemispheres having a positive impact on the populations of its predators.

Additional trophic ecology studies in the Pacific Ocean are necessary to identify the variation in the use of food sources by *I. oxyrinchus*. There are few studies on this shark in the north and south Pacific, and this is the first report with a full description of the feeding habits of this shark in the central Pacific. These results and those reported for the same species in the Atlantic Ocean increase our understanding of the importance of this species in the ecosystem. We suggest integrating this information into ecological models for the better management of the shark population.

Acknowledgments

The authors thank the University Laica "Eloy Alfaro" of Manabí and its Central Department of Research for their support through the project "Trophic Ecology of the Large Pelagic Species of the Ecuadorian Pacific 2014-2015", of which this manuscript is a part. The authors would like to thank all the students who directly contributed to this project for their help collecting samples. An additional thanks to the fishermen of Santa Rosa, Salinas for granting us access to the sharks. Thanks to Sarah Young for her support and observations on the English version.

References

Abascal, F. J., Quintans, M., Ramos-Cartelle, A. and Mejuto, J. (2011). Movements and environmental preferences of the shortfin mako, *Isurus oxyrinchus*, in the southeastern Pacific Ocean. *Mar. Biol.*, *158*, 1175-1184.

Aguiar, A. A. and Valentin, J. L. (2010). Biologia e ecologia alimentar de elasmobrânquios (Chondrichthyes: Elasmobranchii): uma revisão dos métodos e do estado da arte no Brasil. *Oecologia Australis*, *14*, 464-489.

Alegre, A., Ménard, F., Tafur, R., Espinoza, P., Argüelles, J., Maechara, V., Flores, O., Simier, M. and Bertrand, A. (2014). Comprehensive model of jumbo squid *Dosidicus gigas* trophic ecology in the Northern Humboldt Current System. *PLoS One.*, *9*, e85919, http://dx.doi.org/10.1371/journal.pone.0085919.

Allen, G. R. and Robertson, D. R. (1998). *Peces del Pacífico Oriental Tropical* (First edition). México. Crawford House Press Pty Ltd, Mexico.

Amundsen, P. A., Gabler, H. M. and Staldvik, F. J. (1996). A new approach to graphical analysis of feeding strategy from stomach contents data-modification of the Costello (1990) method. *J. Fis. Biol.*, *48*, 607-614.

Aurioles-Gamboa, D. (1992). Inshore-ofshore movements of pelagic red crabs *Pleuroncodes planipes* (Decapoda, Anomura, Galatheidae) off the Pacific coast of Baja California Sur, Mexico. *Crustacea*, *62*, 71-84.

Barker, M. J. and Schluessel, V. (2005). Managing global shark fisheries: suggestions for prioritizing management strategies. *Aquat. Conserv.*, *15*, 325-347.

Bazzino, G., Gilly, W. F., Markaida, U., Salinas-Zavala, C. A. and Ramos-Castillejos, J. (2010). Horizontal movements, vertical-habitat utilization and diet of the jumbo squid (*Dosidicus gigas*) in the Pacific Ocean off Baja California Sur, Mexico. *Prog. Oceanogr.*, *86*, 59-71.

Biton-Porsmoguer, S., Banaru, D., Boudouresque, C. F., Dekeyser, I., Viricel, A. and Merchán, M. (2015). DNA evidence of the consumption of short-beaked common dolphin *Delphinus delphis* by the shortfin mako shark *Isurus oxyrinchus*. *Mar. Ecol-Prog. Ser.*, *532*, 177-183.

Bizzarro, J. J., Robinson, H. J., Rinewalt, C. S. and Ebert, D. A. (2007). Comparative feeding ecology of four sympatric skate species off central California, USA. *Environ. Biol. Fish.*, *80*, 197-220.

Bornatowski, H., Wedekin, L. L., Heithaus, M. R., Marcondes, M. C. C. and Rossi-Santos, M. R. (2012). Shark scavenging and predation on cetaceans at Abrolhos Bank, Eastern Brazil. *J. Mar. Biol. Assoc. UK*, *928*, 1767-1772.

Bornatowski, H., Navia, A. F., Braga, R. R., Abilhoa, V. and Maia, C. M. F. (2014). Ecological importance of sharks and rays in a structural food web analysis in southern Brazil. *ICES J. Mar. Sci.*, 1-7. doi:10.1093/icesjms/fsu025.

Bress, M. (1993). The behavior of sharks. *Rev. Fish Biol. Fisher.*, *3*, 133-159.

Cailliet, G. M. (1976). Several approaches to the feeding ecology of fishes. In: Simenstad, C. A., Lipovsky, S. J. (Eds.), *Fish Food Habits Studies*, 1st Pacific Northwest Technical Workshop Proceedings. Astoria, OR, October 13-15. University of Washington, Washington Sea-Grant Publications, Seattle. 1-13.

Chang, J. H. and Liu, K. M. (2009). Stock assessment of the shortfin mako shark (*Isurus oxyrinchus*) in the Northwest Pacific Ocean using per recruit and virtual population analyses. *Fish. Res.*, *98*, 92-101.

Clarke, M. R. (1986). *A handbook for the identification of Cephalopod beaks* (First edition). Oxford University Press. New York.

Clarke, M. R. (1996). Cephalopods as prey. III. Cetaceans. *Philos. T. Roy. Soc. B.*, *351*, 1053-1065.

Cliff, G. and Dudley S. F. J. (1991). Sharks caught in the protective gill nets off Natal South Africa. 4. The bull sharks *Carcharhinus leucas* Valenciennes. *S. Afr. J. Marine Sci.*, *10*, 253-270.

Cliff, G., Dudley, S. F. J. and Davis, B. (1990). Sharks caught in the protective gill nets off Natal, South Africa. 3. The shortfin mako shark *Isurus oxyrinchus* (Rafinesque). *S. Afr. J. Marine Sci.*, *91*, 115-126.

Clothier, C. R. (1950). A Key to Some Southern California Fishes Based on Vertebral Characters. *Cal. Fish. Game.*, *79*, 1-83.

Compagno, L. J. V. (2001). Sharks of the World. *An annotated and illustrated catalogue of Sharks species known to date. Vol. 2, Bullhead, mackerel and carpet sharks (Heterodontiformes, Lamniformes and Orectolobiformes).* Species Catalogue for Fishery. Purposes no. 1. FAO. Rome.

Compagno, L. J. V., Dando, M. and Fowler, S. (2005). *Guía de campo de los tiburones del mundo* (First edition). Barcelona, Ediciones Omega S. A.

Consoli, P., Romeo, T., Battaglia, P., Castriota, L., Esposito, V. and Andaloro, F. (2008). Feeding habits of the albacore tuna *Thunnus alalunga* (Perciformes, Scombridae) from Central Mediterranean Sea. *Mar. Biol.*, *155*, 113-120.

Cortés, E. (1999). Standardized diet compositions and trophic levels of sharks. *ICES J. Mar Sci.*, *56*, 707-717.

Cortés, E. (2000). Life History Patterns and Correlations in Sharks. *Rev. Fish. Sci.*, *8*, 299-344.

Dulvy, N. K., Baum, J. K., Clarke, S., Compagno, L. J. V., Cortés, E., Domingo, A, Fordham, S., Fowler, S., Francis, M. P., Gibson, C., Martínez, J., Musick, J. A., Soldo, A., Stevens, J. D. and Valenti, S. (2008). You can swim but you can't hide: the global status and conservation of oceanic pelagic sharks and rays. *Aquat. Conserv.*, *18*, 459-482.

Dulvy, N. K., Fowler, S. L., Musick, J. A., Cavanagh, R. D., Kyne, P. M., Harrison, L. R., Carlson, J. K., Davidson, L. N. K., Fordham, S. V., Francis, M. P., Pollock, C. M., Simpfendorfer, C. A., Burgess, G. H., Carpenter, K. E., Compagno, L. J. V., Ebert, D. A., Gibson, C., Heupel, M. R., Livingstone, S. R., Sanciangco, J. C., Stevens, J. D., Valenti, S. and White, W. T. (2014). Extinction risk and conservation of the world's sharks and rays. *eLife*, 1-34. DOI: http://dx.doi.org/10.7554/eLife.00590.

Ebert, D. A. and Stehmann. M. F. W. (2013). *Sharks, Batoids and Chimeras of the North Atlantic. Species Catalogue for Fishery. Purposes no. 7.* Rome FAO.

Estrada, J. A., Rice, A. N., Lutcavage, M. E. and Skomal, G. B. (2003). Predicting trophic position in sharks of the north-west Atlantic Ocean using stable isotope analysis. *J. Mar. Biol. Assoc. UK.*, *83*, 1347-1350.

Food and Agriculture Organization of the United Nations (FAO). (2015). *Isurus oxyrinchus* (Rafinesque, 1809). Food and Agriculture Organization of the United Nations Species Fact Sheets, 1-6.

Ferretti, F., Myers, R. A., Serena, F. and Lotze, H. K. 2008. Loss of large predatory sharks from the Mediterranean Sea. *Conserv. Biol*, *22*, 952-964.

Ferretti, F., Worm, B., Britten, G. L., Heithaus, M. R. and Lotze, H. K. (2010). Patterns and ecosystem consequences of shark declines in the ocean. *Ecol. Lett.*, *13*, 1055-1071. DOI: 10.1111/j.1461-0248.2010.01489.x.

Field, J., Baltz, K., Philips, J. and Walker, W. (2007). Range expansion and trophic interactions of the jumbo squid *Dosidicus gigas*, in the California Current. *Cal. Coop. Ocean. Fish.*, *48*, 131–146.

Fischer, W., Krupp, F., Schneider, W., Sommer, C., Carpenter, K. E. and Niem, V. H. (1995a). Guía FAO para la identificación de especies para los fines de la pesca. *Pacífico Centro-Oriental.*, Vol. *2*. Vertebrados-Parte 1. Rome, FAO.

Fischer, W., Krupp, F., Schneider, W., Sommer, C., Carpenter, K. E. and Niem, V. H. (1995b). Guía FAO para la identificación de especies para los fines de la pesca. *Pacífico Centro-Oriental.*, Vol. *3*. Vertebrados-Parte 2. Rome, FAO.

Fowler, S. (2014). The conservation status of migratory sharks. Bonn Germany, UNEP/CMS Secretariat. (http://sharksmou.org/cms-publications).

Frisk, M. G., Miller, T. J. and Fogarty, M. J. (2001). Estimation and analysis of biological parameters in elasmobranch fishes: a comparative life history study. *Can. J. Fish. Aquat. Sci.*, *58*, 969-981.

Frisk, M. G., Miller, T. J. and Dulvy, N. K. (2005). Life histories and vulnerability to exploitation of Elasmobranchs: Inferences from elasticity, perturbation and phylogenetic analyses. *J. Northw. Atl. Fish. Sci.*, *35*, 27-45.

Galván-Magaña, F., Polo-Silva, C., Hernández-Aguilar, S. B., Sandoval-Londoño, A., Ochoa-Díaz, M. R., Aguilar-Castro, N., Castañeda-Suárez, D., Chavez-Costa, A. C., Baigorrí-Santacruz, A., Torres-Rojas, Y. E. and Abitia-Cárdenas, L. A. (2013). Shark predation on cephalopods in the Mexican and Ecuadorian Pacific Ocean. *Deep-Sea Res. Pt. II*, *95*, 52-62.

García-Godos, I. (2001). Patrones morfológicos del otolito sagitta de algunos peces óseos del mar peruano. B. *Instituto del Mar de Perú*, *20*, 1-83.

Gorni, G. R., Goitein, R. and Amorim, A. F. (2013). Description of diet of pelagic fish in the southwestern Atlantic, Brazil. *Biota Neotrop*, *13*, 61-69.

Gorni, G. R., Loibel, S., Goitein, R. and Amorim, A. F. (2012). Stomach contents analysis of shortfin mako (*Isurus oxyrinchus*) caught off southern Brazil: A Bayesian Analysis. *Collective Volume of Scientific Papers-ICCAT*, *68(5)*, 1933-1937.

Heithaus, M. R. (2002). Predator-prey and competitive interactions between sharks (Order: Selachii) and dolphins (Suborder: Odontoceti): a review. *J. Zool.*, *253*, 53-68.

Heithaus, M. R. (2005). Habitat use and group size of pied cormorants (*Phalacrocorax varius*) in a seagrass ecosystem: possible effects of food abundance and predation risk. *Mar. Biol.*, *147*, 27-35.

Heupel, M. R., Knip, D. M., Simpfendorfer, C. A. and Dulvy, N. K. 2014. Sizing up the ecological role of sharks as predators. *Mar. Ecol-Prog. Ser.*, *495*, 291-298.

Hurlbert, S. H. (1978). The measurement of niche overlap and some relatives. *Ecology*, *59*, 67-77.

Hurtubia, J. (1973). Trophic diversity measurement in sympatric predatory species. *Ecology*, *54*, 885-890.

Hussey, N. E., Brush, J., McCarthy, I. D. and Fisk, A. T. (2010). δ15N and δ13C diet–tissue discrimination factors for large sharks under semi-controlled conditions. *Comp. Biochem. Phys. A.*, *155*, 445–453.

Hussey, N. E., MacNeil, M. A., McMeans, B. C., Olin, J. A., Dudley, S. F. J., Cliff, G., Wintner, S. P., Fennessy, S. T. and Fisk A. T. (2014). Rescaling the trophic structure of marine webs. *Ecol. Lett.*, *17*, 239-250.

Jennings, S. and Collingridge, K. (2015). Predicting Consumer Biomass, Size-Structure, Production, Catch Potential, Responses to Fishing and Associated Uncertainties in the World's Marine Ecosystems. *PLoS ONE*, *10*, 1-28.

Jered, P. and Roper, C. F. E. (2010). Cephalopods of the world. An annotated and illustrated catalogue of cephalopod species known to date. Myopsid and Oegopsid Squids. *Species Catalogue for Fishery Purposes no. 4.* Vol. 2. Rome, FAO.

Krebs, C. J. (1999). *Ecological methodology*. California, Addison Wesley Longman.

Langton, R. W. (1982). Diet overlap between the Atlantic cod *Gadus morhua*, silver hake *Merluccius bilinearis* and fifteen other northwest Atlantic finfish. *Fish. Bull.*, *80*, 745-759.

Loefer, J. K., Sedberry, G. R. and McGovern, J. C. (2005). Vertical movements of a Shortfin mako in the Western North Atlantic as determined by Pop-up satellite tagging. *Southeastern Nat.*, *4*, 237-246.

Lopez, K. S., Meléndez, C. R. and Barría, M. P. (2009). Alimentación del tiburón marrajo *Isurus oxyrinchus* Rafinesque, 1810 (Lamniformes: Lamnidae) en el Pacífico Suroriental. Rev. Biol. Mar. *Oceanog.*, *44*, 439-451.

Lopez, K. S., Barría, M. P. and Meléndez, C. R. (2012). Feeding and trophic relationship of two highly migratory sharks in the eastern South Pacific Ocean. *Pan-Am. J. Aquat. Sci.*, *7(1)*, 50-56.

Lowe, C. G., Wetherbee, B. M., Crow, G. L. and Tester, A. L. (1996). Ontogenetic dietary shifts and feeding behavior of the tiger shark, *Galeocerdo cuvier*, in Hawaiian waters. *Environ. Biol. Fish.*, *47*, 203-211.

Madigan, D. J., Carlisle, A. B., Dewar, H., Snodgrass, O. E., Litvin, S. Y., Micheli, F. and Block, B. A., (2012). Stable isotope analysis challenges wasp-waist food web assumptions in an upwelling pelagic ecosystem. *Sci. Rep. 2*, e654. http://dx.doi.org/10.1038/srep00654.

Maia, A., Queiroz, N., Correira, J. P. and Cabral, H. (2006). Food habits of the Shortfin mako, *Isurus oxyrinchus*, off the southwest coast of Portugal. *Environ. Biol. Fish.*, *77*, 157-167.

Malpica-Cruz, L., Herzka, S. Z., Sosa-Nishizaki, O. and Escobedo-Olvera, M. A. (2013). Tissue-specific stable isotope ratios of shortfin mako (*Isurus oxyrinchus*) and white (Carcharodon carcharias) sharks as indicators of size-based differences in foraging habitat and trophic level. *Fish. Oceanogr.*, *22*, 429-532.

Marcaida, U. and Gilly, W. F. (2016). Cephalopods of Pacific Latin America. *Fish. Res.*, *173*, 113-121.

Maya-Meneses, C. I., Torres-Rojas, Y. E., Galván-Magaña, F., Aguiñiga-García, S. and Trasviña-Carrillo, L. D. (2016). Trophic overlap between blue sharks (*Prionace glauca*) and shortfin makos (*Isurus oxyrinchus*): Trophic linkages between two shark species in the Eastern Pacific Ocean food web. *Food Webs. 7*, 13-19.

Monteiro, M. S., Vaske-Júnior, T., Barbosa, T. M. and Alves, M. D. O. (2006). Predation by a shortfin mako shark, *Isurus oxyrinchus*, Rafinesque, 1810, on a pantropical spotted dolphin, *Stenella atenuatta*, calf in central Atlantic waters. *Latin Am. J. Aquatic Mammals*, *5*, 141-144.

Motta, P. J. and Wilga, C. D. (2001). Advances in the study of feeding behaviors, mechanisms, and mechanics of Sharks. *Environ. Biol. Fish.*, *60*, 131-156.

Mucientes, G. R., Queiroz, N., Sousa, L. L., Tarroso, P. and Sims, D. W. (2009). Sexual segregation of pelagic sharks and the potential threat from fisheries. *Biol. Lett.*, *5*, 156-159.

Mucientes-Sandoval, G. R. and Saborido-Rey, F. (2008). Acercamiento a la composición de la dieta de *Isurus oxyrinchus* Rafinesque, 1810 (Elasmobranchii: Lamnidae) en aguas internacionales del Pacífico Sur Central. *Rev. Invest. Mar.*, *29*, 145-150.

Navia, A. F., Cortés, E. and Mejía-Falla, P. A. (2010). Topological analysis of the ecological importance of elasmobranch fishes: A food web study on the Gulf of Tortugas, Colombia. *Ecol. Mod.*, *221*, 2918-2926.

Nelson, J. S. (2006). *Fishes of the world* (Fourth Edition). United States of America, John Wiley and Sons.

Newton, K. C., Wraith, J. and Dickson, K. A. (2015). Digestive enzyme activities are higher in the shortfin mako shark, *Isurus oxyrinchus*, than in ectothermic sharks as a result of visceral endothermy. *Fish. Physiol. Biochem*, *41*, 887–898.

Nigmatullin, Ch. M., Nesis, K. N. and Arkhipkin, A. I. (2001). A review of biology of the jumbo squid *Dosidicus gigas* (Cephalopoda: Ommastrephidae). *Fish. Res.*, *54*, 9-19.

North, A. W., Burchett, M. S., Gilbert, C. J. and White, M. G. (1984). Identification of fish from the southern ocean by means of otoliths. *Brit. Antartic Surv. B.*, *62*, 83-94.

Oksanen, J., Blanchet, F., Kindt, R., Legendre, P., O'Hara, R. G., Simpson, G., Solymos, P., Stevens, M. H. and Wagner, H. (2010). Vegan: community ecology package. R package version 1.17-0 Accessible online: http://CRAN.R-project.org/package0vegan.

Pinkas, L., Oliphant, M. S. and Iverson, K. (1971). Food habits of albacore, bluefin tuna, and bonito in California waters. *Calif. Fish. Game.*, *152*, 1-105.

Polidoro, B. A., Brooks, T., Carpenter, K. E., Edgar, G. J. and Henderson, S. (2012). Patterns of extinction risk and threat for marine vertebrates and habitat-forming species in the Tropical Eastern Pacific. *Mar. Ecol. Prog. Ser.*, *448*, 93-104.

Preti, A., Soykan, C. U., Dewar, H., Wells, D. R. J., Spear, N. and Kohin, S. (2012). Comparative feeding ecology of shortfin mako, blue and thresher sharks in the California Current. Environ. *Biol. Fish.*, *95*, 127-146.

Punsly, R. G. and Deriso, R. B. (1991). Estimation of the abundance of yellowfin tuna *Thunnus albacares*, by age groups and regions within the eastern Pacific Ocean. *Inter-American Tropical Tuna Commission Bulletin*, *20(2)*, 99-131.

R Development Core Team. (2010). R: a language and environment for statistical computing. R foundation for statistical computing. Vienna, Austria. http://www.R-project.org.

Roa, R., Gallardo, V. A., Ernst, B., Baltazar, M., Canete, J. I. and Enriquez-Brionnes, S. (1995). Nursery ground, age structure and abundance of juvenile squat lobster *Pleuroncodes monodon* on the continental shelf off central Chile. M*ar. Ecol. Prog. Ser.*, *116*, 47-54.

Robinson, C. J., Anislado, V. and Lopez, A. (2004). The pelagic red crab (*Pleuroncodes planipes*) related to active upwelling sites in the California Current off the west coast of Baja California. *Deep-Sea Res. Pt. II*, *51*, 753-766.

Rosas-Luis, R., Salinas-Zavala, C., Koch, V., Del Monte Luna, P. and Morales-Zárate, V. (2008). Importance of jumbo squid *Dosidicus gigas* (Orbigny, 1835) in the pelagic ecosystem of the central Gulf of California. *Ecol. Mod.*, *218*, 149–161.

Rosas-Luis, R., Loor-Andrade, P., Carrea-Fernández, M., Pincay-Espinoza, J. E., Vinces-Ortega, J. and Chompoy-Salazar, L. (2015). Cephalopod species in the diet of large pelagic fish (sharks and billfishes) in Ecuadorian waters. *Fish. Res.*, *173*, 159-168.

Smith, P. E. and Zaret, M. T. (1982). Bias in estimating niche overlap. *Ecology*, *63*, 1248-1253.

Stillwell, C. E. and Kohler, N. E. (1982). Food, Feeding habits, and Estimates of Daily Ration of the Shortfin Mako (*Isurus oxyrinchus*) in the Northwest Atlantic. Can. *J. Fish. Aquat. Sci.*, *39(3)*, 407-414.

Vaske-Júnior, T. and Rincón-Filho, G. (1998). Conteúdo estomacal dos tubarões azul (*Prionace glauca*) e anequin (*Isurus oxyrinchus*) em águas oceânicas no sul do Brasil. *Rev. Brasileira de Biología*, *58*, 445-452.

Velasco-Tarelo, P. M. (2005). Hábitos alimenticios e isótopos de ^{13}C y ^{15}N del tiburón mako *Isurus oxyrinchus* (Rafinesque, 1810) en la Costa occidental de Baja California Sur. Master Thesis. Centro Interdisciplinario de Ciencias Marinas. México.

Vetter, R., Kohin, S., Preti, A., McClatchie S. and Dewar, H. (2008). Predatory interactions and niche overlap between mako shark, *Isurus oxyrinchus*, and jumbo squid, *Dosidicus gigas*, in the California Current. *Cal. Coop. Ocean. Fish.*, *49*, 142-156.

Wolff, G. A. (1984). Identification and estimation of size from the beaks of 18 species of cephalopods from the Pacific Ocean. NOAA *Technical Report NMFS*, *17*, 1-50.

Wood, A. D., Wetherbee, B. M., Francis, J., Kohler, N. E. and Wilga, C. (2009). Recalculated diet and daily ration of the shortfin mako (*Isurus oxyrinchus*), with a focus on quantifying predation on bluefish (*Pomatomus saltatrix*) in the northwest Atlantic Ocean. *Fish. Bull.*, *107*, 76-88.

Xavier, J. C. and Cherel, Y. (2009). *Cephalopod beak guide for the southern ocean* (First edition). Cambridge. Brit. Atartic Surv. B.

Yatsu, A. (1995). The role of slender tuna, *Allothunnus fallai*, in the pelagic ecosystems of the South Pacific Ocean. *Jpn. J. Ichthyol.*, *41*, 367–377.

In: Advances in Marine Biology. Volume 1 ISBN: 978-1-53610-131-7
Editors: Adam Kovács and Patrik Nagy © 2016 Nova Science Publishers, Inc.

Chapter 7

ECOLOGICAL INTERDEPENDENCE IN MARINE HABITATS: BIO-ECONOMIC MANAGEMENT OF A SPANISH MIXED FISHERY

M. Dolores Garza-Gil**, *Marcos I. Pérez-Pérez, Manuel M. Varela-Lafuente and Juan C. Surís-Regueiro
Department of Applied Economics, University of Vigo, Vigo, Spain

ABSTRACT

Ecological interactions amongst fish stocks can play a dominant role in how marine ecosystems are structured, and they can have a significant impact on the populations' growth dynamics. The analytical framework for proposing management guidelines has been developed based on bio-economic models that only account for a single species, both in terms of biological and well as economic aspects, and this diminishes the importance of interactions among the various marine community components. Bearing in mind the implications of the ecological dependencies that are created in multi-species fisheries, that bio-economic model can be extended. A better understanding of trophic relationships among fish stocks enables the development of management models for multi-species fisheries—that is, for most European fisheries. This chapter examines a fishery encompassing two of the Spanish fleet's

* Corresponding Author Email: dgarza@uvigo.es.

target species that exhibit a significant ecological predatory interdependence: the blue whiting and the southern stock of European hake. Such interactions should be considered when establishing the Total Allowable Catch (TAC) because the catch level for one species will alter populations of the other, thereby affecting long-run sustainability of commercial fishing and of the marine environment. Within this fishery, both the catch and the spawning biomass for each species are trending downward. The trophic interaction between predator (hake) and prey (blue whiting) is included into the management problem of jointly determining fishing quotas for each species. The estimations indicate that increases in blue whiting biomass (prey) have generated a positive effect on hake biomass along that period 1988-2014, it is represented by a plus sign in the parameter for interdependence in the hake growth function. However, increases in hake biomass (predator) lead to a reduction in blue whiting biomass and this is represented by a minus sign in the parameter for interdependence in the blue whiting growth function. These results are consistent with a relationship of interdependence predator-prey. Incorporating this interaction into stock dynamics for both species and applying optimum control theory, we obtain the optimal spawning stock biomass values for both species and the corresponding catch values. The results show catch levels are lower than prevailing EU-determined levels for both species. In addition, as the social discount rate increases and, therefore society's preferences in the near future, the biomass of both species decreases, while catches or potential landings increase.

Keywords: mixed fisheries management, predator-prey model, European fisheries, Southern hake, blue whiting

INTRODUCTION

The effects of fishing are reflected not only in the target species but also in the entire ecosystem that inhabit. There are scientific evidences that it is occurring in the marine ecosystems and this imposes to find alternative approaches to the current fisheries management scheme, so it has been expressed the need that fisheries research and management to move towards an ecosystem-based approach [1]. The introduction of ecological considerations in fisheries management looks like this as a strategy to ensure the sustainability of fisheries and marine ecosystems. More and more scientists are recommending multi-species approaches to fishery management, due to the complex series of interactions between a fishery's different species [1, 2]. And in practice, fisheries are exploited as a multi-species resource [3]. Biological interactions amongst fish stocks can play a dominant role in how marine ecosystems are structured, and they can have a significant impact on the fish populations' dynamics [2, 4]. More specifically, predation and cannibalism are sources of mortality that can regulate a population's size and recruiting [2, 5-7]. Therefore, fishery management can largely benefit from better comprehension of the role of biological interactions—including trophic relationships—in the resource's dynamics [8-26].

Within the context of ecosystem management, the multispecies approach occupies an important place in fisheries management, since biological species interactions regulate the recruitment and the size of the populations of the ecosystem. The study of trophic relationships as sources of mortality may improve the understanding of the population dynamics and the management of marine resources. Fish stock assessment methods have evolved from monospecific models, studying the population of a single species whose dynamics is affected only by its own biomass and the effect of fishing on the same, towards multispecies models. The multispecies models take into account ecological or technological interactions to optimize yields form the ecosystem as a whole. The multispecies models are in many cases extensions of monospecific models, as it is the case of the excess production models, aggregate production models, multispecies per recruit models, virtual population analysis model, and food and trophic network models. Ecological interdependence occurs when there is a certain relationship between two exploited stocks [16, 17] or between two different population segments of the same stock. From there, we can go on to define two types of ecological interactions: competition and predation. Competition is defined according to the negative effects that one organism exerts over another by consuming or interfering with its access to a resource that has a limited availability (supply) [18]. This interaction leads to a reduction in the survival, growth, and/or reproduction of competitors, particularly those that, hierarchically speaking, are subordinate [19, 20]. And predation is the consumption of one organism or species (prey) by another organism/species (predator) through carnivory, herbivory or omnivory [20]. On other hand, technological interaction occurs when two or more fishing gears are operating on the same population of fish.

Multispecies bioeconomic models arise from the extension of Gordon-Schaefer model for a single target stock, incorporating logistic equations predator-prey. Predator-prey models constitute a particular case within the multispecies models. They are based on the equations of Lotka-Volterra, describing the trophic interactions between the predator and the prey. The predator-prey basic model assumes exponential or Malthusian growth for the population of prey in absence of predators. This model incorporates the intrinsic growth rates of the predator and the prey and the carrying capacity of the ecosystem to each of the two species. The predator-prey model with capture poses a situation of commercial exploitation of the species and takes into account the discount rate of the fishery. This model aims to maximize the present value of the net benefit from the exploitation of both species: predator and prey.

This chapter deals with the management of a multispecies fishery (mixed fishery) with two target stocks that exhibit significant trophic interaction (ecological interdependence) as well as predation: the southern stock of European hake (*Merluccius merluccius*) and the blue whiting stock (*Micromesistius poutassou*). At this North Atlantic fishery, Southern hake is the predator and blue whiting the prey.

THE EUROPEAN HAKE AND BLUE WHITING MIXED FISHERY

The European hake and blue whiting are two target species for the Spanish longline and trawl fleet in Iberian Atlantic waters, although the latter is also used for horse mackerel, Atlantic mackerel, and octopus. Both species are primarily distributed across the continental shelf, where they spawn during the winter months.

European hake is widely distributed throughout the north-eastern Atlantic Ocean [27]. It is a demersal benthopelagic species that can reach depths that vary from 30 to 1000 meters, although they generally dwell at depths ranging between 70 to 370 meters. They form schools that stick close to the coastline in the summer and keep a greater distance during the winter. They reproduce at depths of 100 to 300 meters. Egg-laying season runs from January to May in the Bay of Biscay. Juveniles live in the muddy depths until about three years of age, at which point they move closer to the coast. Adults can typically be found on the continental slope. The European hake is a predator at the top of the northeast Atlantic demersal trophic pyramid. It primarily feeds on other species of fish like the anchovies (*Engraulis encrasicholus*), sardines (*Sardina pilchardus*), and blue whiting (*Micromesistius poutassou*). According to the organization that analyses the fish stocks in the Northeast Atlantic grounds (ICES), there are two stock units for this species: the southern stock, in zones VIIIc and IXa, and the northern stock, in the remaining NE Atlantic zones.

Table 1. Hake biomass and landings in in ICES zones VIIIc and IXa. 1988-2014

	Spanish landings (tons)	Total landings (tons)	SSB (tons)
1988	10600	16700	27035
1989	9200	13800	19905
1990	9800	13200	16287
1991	8900	12800	16462
1992	8840	13800	15527
1993	7840	11500	12775
1994	6800	9900	8908
1995	8600	12200	7104
1996	7000	9700	8552
1997	6100	8500	6557
1998	5000	7700	5812
1999	4000	7200	7545
2000	4700	7900	8823
2001	4400	7580	8973
2002	3900	6700	9379
2003	4700	6740	9075
2004	4900	6940	8962
2005	6200	8300	9286
2006	8700	10800	10816
2007	12800	14930	12786
2008	14600	16770	12695
2009	16900	19240	13830
2010	13000	15740	12940
2011	6500	17062	15229
2012	5800	14573	14756
2013	7200	13231	14036
2014	7300	12011	18836

Source: Authors compilation from International Council for Exploration of the Sea (ICES) [27].

Blue whiting is a demersal species from the gadidae family, and it can be found throughout the northern Atlantic [28]. Its habitat is the ocean, and as a benthopelagic species, it lives mainly along the continental shelf and slope, distributed vertically at depths of between 150 and 1000 meters, though it is most commonly found at 300-400 meters below the surface. Blue whiting spawns between February and June and reaches first maturity at three years of age. Growth is rapid, and females are typically larger than males. The blue whiting's diet consists primarily of various types of crustaceans.

A number of studies have confirmed the ecological interdependence of the hake and blue whiting populations in Iberian waters analyzing the diet of the European hake in the Cantabrian Sea (zone VIIIc), and based on a seasonal analysis of the stomach contents of the fish sampled [29-31]. They highlight the importance of the blue whiting as the hake's primary prey, particularly amongst the adult population, while most juveniles consume cupleoids, at more shallow depths, since at such depths there are fewer blue whiting. Other studies confirm the existence of a significant relationship between the size of the hake and the size of the blue whiting using statistical regression analysis [9, 32]. Likewise, hake is far more dependent upon blue whiting in Iberian fishing grounds (zone VIIIc) than in the northern part of the Bay of Biscay and the Celtic Sea [2, 9, 32]. Other authors studied the diet of the blue whiting, hake, horse mackerel and Atlantic mackerel in Portuguese waters (Zone IXa) and analyzed their stomach contents [8], and they concluded that in terms of percentage, occurrence, and weight, the blue whiting is the hake's most important prey.

Hake landings and biomass in ICES zones VIIIc and IXa are shown in Table 1 for period 1988-2014. These variables have changed over the past 25 years in zones VIIIc and IXa, as Figure 1 demonstrates. The landings, in general, exhibit a downward trend throughout most of this period, both in terms of the total catch for the industry overall as well as the Spanish fleet. However, beginning in 2007 there was significant recovery, though the numbers once again declined at the end of the period. Spawning stock biomass (SSB) also exhibited a downward trend up until 2006, at which point a notable recovery was made. Note that between 1994 and 1998, the landings surpassed the reproductive biomass; this scenario surfaced once again between 2007 and 2010.

As a result of the elevated hake landings, the species has been outside biosafety level limits since the 1990s [27]. This situation has created an incentive for implementing a recovery plan for the species aimed at recovering the stock and bringing it back within the established biosafety level limits, that is 35,000 tons of reproductive stock biomass (European Council-EC Regulation 2166/2005). According to that plan, in 2006 the European Commission (EC) applied an annual reduction of 10% to the total allowable catch (TAC) and fishing effort for the fishing fleets (measured as the sum of the engine power in kilowatt and the number of days fishing in the area). Over the last few years, the positive effects of this plan have started to become noticeable, to such an extent that the EC began increasing the TAC for this species in 2011.

Table 2. Blue whiting biomass and landings in in ICES zones VIIIc and IXa. 1988-2014

	Spanish landings (tons)	Total landings (tons)	SSB (tons)
1988	24847	30826	177889
1989	30108	33665	170549
1990	29490	32354	154157
1991	19180	31993	197768
1992	23794	28722	264631
1993	31020	32256	257187
1994	28118	29468	249801
1995	25379	27664	231837
1996	21538	25099	216611
1997	27683	30122	224706
1998	27490	29390	322130
1999	23777	26402	390036
2000	22622	24654	422364
2001	23218	24964	463867
2002	17506	19165	565858
2003	13825	16476	697320
2004	15612	19549	700133
2005	17643	22833	655494
2006	15173	20496	649828
2007	13557	17454	541748
2008	14342	18562	420634
2009	20637	22680	324617
2010	12891	14373	304926
2011	2416	3019	321000
2012	6726	8681	339600
2013	15274	17330	391800
2014	32065	34215	396500

Source: Authors compilation from International Council for Exploration of the Sea (ICES) [28].

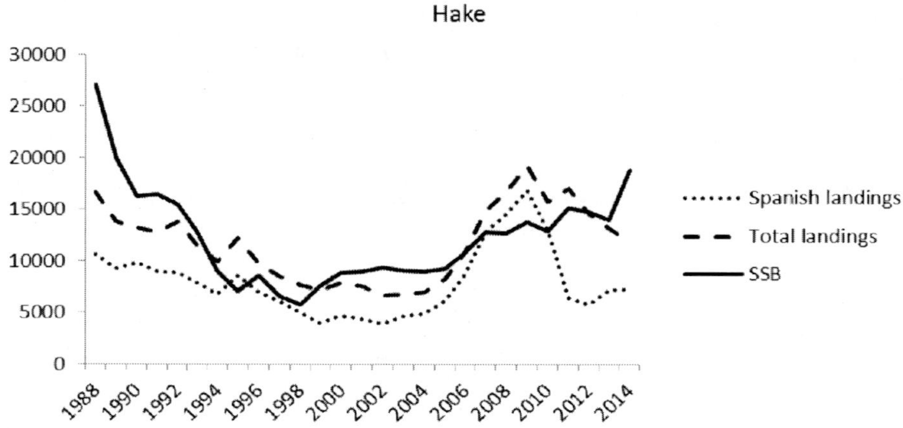

Source: Authors compilation from Table 1.

Figure 1. Hake landings (tons) and estimated SSB (tons) in ICES zones VIIIc and IXa. 1988-2014.

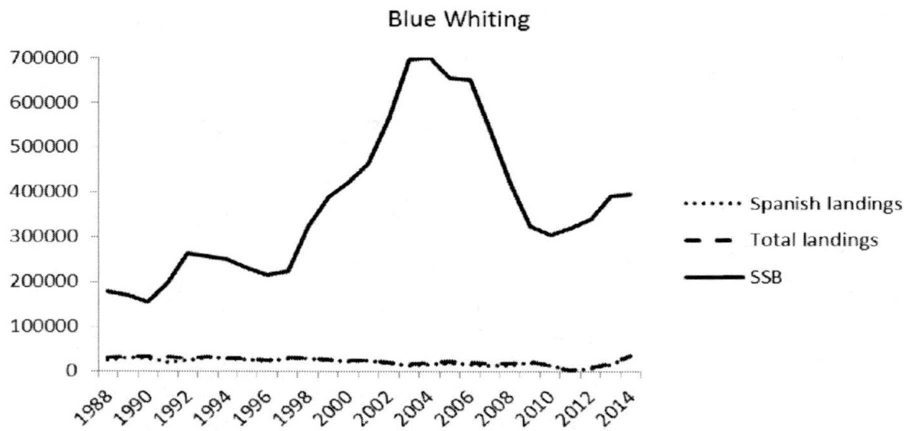

Source: Authors compilation from Table 2.

Figure 2. Blue whiting landings (tons) in ICES zones VIIIc and IXa and estimated SSB (10 tons). 1988–2014.

As for the blue whiting, Table 2 shows the biomass and landings for this species in ICES zones VIIIc and IXa for period 1988-2014. The evolution of these variables is showed more clearly in Figure 2. Spanish landings in zones VIIIc and IXa have stayed relatively stable, at around twenty thousand tons since 2010, at which time a considerable decline was noted, due to the decreased TAC established for this species in 2010. The SSB demonstrates a sustained increase until the end of the 1990s, where a significant decrease is observed. Nevertheless, this variable shows signs of recovery in the last few years due to a drop in catches over the past decade, in combination with an increase in recruitment since 2009. It is estimated that in 2013, the SSB was already above the precautionary limits [28]. This has allowed European Council agreed a TAC higher in 2014 and 2015 than in previous years.

The Spanish fishing fleet has also exhibited a downward trend over the past 25 years, going from 279 ships in 1988 to 180 in 2013 [28, 33]. As far as economic data for the fishing grounds is concerned, Table 1 shows a standard ship's average revenue and costs between 2010 and 2014. The largest segment of the costs corresponds to personnel expenses (48% of the ship's average revenue), while the gross cash flow is positive and accounts for approximately 18% of the total revenue.

Table 3. Average technical characteristics and economic figures for Spanish fishery vessels

Variable	Average 2010–2014
Number of vessels	180
Crew	11
Length (m)	28.4
Power (kW)	466.29
Gross tonnage (GT)	224.2
Income (€)	514,414.17
Crew share (€)	248,101.95
Other vessel costs (€)	176,186.85
Gross cash flow / mixed incomes (€)	90,125.36
Landings of hake per vessel (tons) [a]	19.95
Landings of blue whiting per vessel (tons) [a]	83.02
Unit price for hake landings (€/kg) [a]	4.11
Unit price for blue whiting landings (€/kg) [a]	1.87
Unit cost for hake landings (€/kg) [a]	2.87
Unit cost for blue whiting landings (€/kg) [a]	1.06

[a] Figures estimated from the landings of the fleet in the port of Vigo.
Notes: Monetary values in €2014. kW = kilowatts; m = meters.
Source: Own compilation from [33, 38, 39].

Lastly, Table 3 also shows the technical characteristics and the main economic data for this Spanish fleet and, in particular, the unit costs and prices per landed tons for each species. Data are shown as an average for the period 2010-2014. Since we do not know the composition of the catches for the entire Spanish fleet in this fishery, the costs have been determined based on a representative sample. More specifically, the unit costs have been estimated using the weight that corresponds to each species (13.5% hake and 20.8% blue whiting) in the landings of vessels based in the Port of Vigo in 2010-2014 (these ships represent 37.8% of the total of the fishery's Spanish fleet), bearing in mind the average tonnage of each species that these ships landed. Additionally, the unit prices were estimated based on prices obtained in the Vigo market for each species.

An Applied Management Model

The ecological interdependence between hake and blue whiting in this fishery, which in this case is a predatory relationship, influences the marine ecosystem structure and trophic chain. Increasing or decreasing the TAC for one of the two species will impact this structure and its sustainability in the long run and, ultimately, the sustainability of the industry in this multi-species fishery. Therefore, it is important to take a simultaneous decision regarding catch levels for both stocks, based on their ecological relationship.

To that end, based on [34], we propose the following net growth functions for each species, taking into account the ecological relationships between hake and blue whiting:

$$\frac{dX}{dt} = r_X\, X\left[1 - \frac{X}{K_X}\right] - H_X + \alpha XY = F(X) - H_X \tag{1}$$

$$\frac{dY}{dt} = r_Y\, Y\left[1 - \frac{Y}{K_Y}\right] - H_Y + \beta XY = G(Y) - H_Y \tag{2}$$

where X stands for hake spawning biomass; Y is for blue whiting spawning biomass; r and K are biological and ecological parameters and denote the intrinsic growth rate and the average load capacity corresponding to each species, respectively; H is for catches (or landings); and α and β denote the coefficient for ecological interaction between hake and blue whiting, respectively. According to the traditional notation in the bio-economic literature on fisheries, the function of net fishery benefits at moment t is determined by the sum of the net benefits generated by the catch for both species for that moment t (most commonly t is considered as a year). Formally, it is defined as the following expression [35]:

$$\pi(t) = (p_X - c_X)H_X + (p_Y - c_Y)H_Y \tag{3}$$

And where p and c stand for the catch unit price and cost of the corresponding species, both parameters are usually assumed constant along the period of study. The problem that the manager or regulator must solve is setting catch levels for each species that will yield greater potential benefits in the long term, given the biological relationships between both species and their net growth limits. Formally, this is expressed as follows [35, 36]:

$$max_{H_X, H_Y} \int_0^{\infty} [(p_X - c_X)H_X + (p_Y - c_Y)H_Y]\, e^{-\delta t} dt \tag{4}$$

$$s.t. \quad \frac{dX}{dt} = F(X) - H_X$$

$$\frac{dY}{dt} = G(Y) - H_Y$$

where δ stands for the social discount rate. This parameter collects the preferences of the society (or country) by the time and the conservation of the living aquatic resources: the higher this parameter, the greater preference for the present moment and greater exploitation of marine resources; and the lower the discount rate, more preference for the future and sustainability of marine biological resources. The yield of government bonds to long-term (ten years) is often used as the discount rate in bio-economic models.

In order to apply the theoretical model to the case study (hake and blue whiting mixed Spanish fishery), we need to know the value of the parameters of the net growth functions in addition to parameters shown in Table 3 (unit price and cost for catches of both species). Econometric regressions were calculated based on the SSB and catch data for the 1988-2014 period gathered by ICES [27, 28] and shown in Tables 1 and 2. The results of these regressions can be seen in Table 4. These estimations indicate that increases in blue whiting biomass (prey) have generated a positive effect on hake biomass along that period 1988-2014 (a plus sign in the parameter for α interdependence in the hake growth function). However, increases in hake biomass (predator) lead to a reduction in blue whiting biomass (a minus sign in the parameter β for interdependence in the blue whiting growth function). These results are consistent with a relationship of interdependence predator-prey; and they are, in general, in line with results from other studies that have also showed this relation analyzing the stomach contents of the fish sampled [2, 8, 9, 29-32].

By adding the value of the parameters shown in Tables 3 and 4 to the above management problem and applying the theory of optimal control [37], we can obtain the optimal SSB values for both species and the corresponding catch values. The results for different discount rate values are shown in Table 5.

As the social discount rate increases—and, therefore, society's preferences in the near future—the biomass of both species decreases, while catches (or potential landings) increase. In 2015, the TAC was set at 13,826 tons for hake and 1,260,000 tons for blue whiting by European Council. These fishing opportunities are slightly higher than they would be if catches for both species were estimated jointly by incorporating their trophic interaction (see Table 5).

Table 4. Statistical results for net growth function of hake and blue whiting

Coefficient	Hake (X)	Blue whiting (Y)
r	1.0983 *** (0.0000)	1.75482 *** (0.0000)
r/K	−3.42881e-0.5 ** (0.0077)	−3.15675e-0.7 *(0.0350)
α	1.9879e-0.7 * (0.0292)	—
β	—	−1.42684e-0.5 *(0.0279)
R^2	0.968498	0.957652

Notes: p-values are reported in parentheses. Coefficient estimates are statistically significant at ***1%, ** 5% or *10% level.

Table 5. Estimated tons of spawning stock biomass (SSB) and total allowable catch (TAC) for different discount rates

Discount rate	SSB		TAC	
	Hake	Blue whiting	Hake	Blue whiting
3%	40499	3016874	10915	670808
4%	38951	2834582	11213	856307
5%	37392	2636035	11344	1020481
6%	35954	2459279	11467	1139988

Conclusion

The rapid growth of the world population is generating a strong pressure on the marine resources as a source of food. Not only overfishing has impact on the target species, but also on the whole of the ecosystem where these species live. It is necessary to use multispecies approaches in fisheries management due to the complex set of interactions among species that occurs in the marine ecosystems. The ecological or biological interactions between species can play a key role in the structuring of the marine communities, and can significantly affect the dynamics of fish populations. Specifically, predation and cannibalism are sources of mortality that can regulate the recruitment and the size of the population. This chapter has developed a multispecies bio-economic model of predator-prey type applied to the mixed fishery for hake and blue whiting in European waters. Both species have a significant trophic interaction, as well as commercial importance for the Spanish fishing fleet which makes them suitable for applying a model of this type.

Hake and blue whiting stocks caught by the Spanish fleet in domestic fishing grounds exhibit significant prey-predator interaction, where hake is the predator and blue whiting the prey. By developing a multi-species bio-economic model for managing both populations in a mixed fishery, instead of contemplating two single-species fisheries, this biological interdependence can be considered as a biological technical externality. When this externality is positive, the optimal stock levels for predator and prey alike are greater than if they were independent populations. More specifically, increases in blue whiting biomass have a positive effect on hake biomass and vice-versa. Similarly, optimal populations of both species increase when the discount rate decreases, while catches would increase if this parameter were increased. The results obtained demonstrate that catches—where the existing trophic interaction is factored into the estimation thereof—are below the TAC levels established by the European Council.

In particular, Spanish fishing boats are currently responsible for 64% and 3% of the TAC established by the European Council for both species in ICES zones VIIIc and IXa. That is, the fleet would bring in approximately 7,000 and 30,000 tons of hake and blue whiting, respectively, if the catches for both stocks were determined in light of their ecological interaction (and while using the 5% discount rate typical of bio-economic studies). However, the TAC established for the Spanish fleet in 2015 was about 9,000 tons of hake and 38,000 tons of blue whiting, or amounts in excess of those that would be allowed had the TAC for both species had been estimated while factoring in their ecological interdependence. Although the 2015 TAC for both species is slightly greater than the fishing opportunities actually obtained for these species, we remark that excess blue whiting catches could well lead to a reduction in the biomass of this species and could therefore—given the species' trophic relationship—also reduce the long-term biomass of hake, which is one of the Spanish fleet's primary target species.

In short, in mixed fisheries where there is an ecological interdependence between species, when the TAC for a given species is estimated on its own, it could have a greater impact on other target species that are not factored into in the management problem. Therefore, it is a good idea to determine the TAC jointly for the species involved and thus internalize the effect of their trophic interaction. This study has provided a very simple example of two stocks, but it is one that is quite relevant to the Spanish fleet, as hake is one of its primary target species. We must continue to gather information on trophic relationships in order to fine tune EU fishery policy recommendations and generate the least impact possible on marine ecosystems, thereby helping to ensure the conservation and sustainability of marine biodiversity as well as sustainability of the fishing activity.

ACKNOWLEDGMENTS

The authors acknowledge financial support from FEDER and Xunta de Galicia (GRC2014/022) and Ministry of Economy and Competitiveness (ECO2014-52412-R and ECO2013-44436-R).

REFERENCES

[1] Botsford, L. W., Castilla, J. C., Peterson C. H. (1997). The management of fisheries and marine ecosystems. *Science*, 277, 509–515.

[2] Mahe, K., Amara, R., Bryckaert, T., Kacher, M., Brylinski, J. M. (2007). Ontogenetic and spatial variation in the diet of hake (Merluccius merluccius) in the Bay of Biscay and the Celtic Sea. *ICES Journal of Marine Science*, 64, 1210–1219.

[3] May, R. M., Beddington, J. R., Clark, C. W., Holt, S. J., Laws, R. M. (1979). Management of multispecies fisheries. *Science*, 205, 267–277.

[4] Rothschild, B. J. (1991). Multispecies interactions on Georges Bank. *ICES Marine Science Symposia*, 193, 86–92.

[5] Neuenfeldt, S., Koster, F. W. (2000). Trophodynamic control on recruitment success in Baltic cod: the influence of cannibalism. *ICES Journal of Marine Science*, 57, 300–309.

[6] Tsou, T. S., Collie, J. S. (2001). Predation-mediated recruitment in the Georges Bank fish community. *ICES Journal of Marine Science*, 58, 994–1001.

[7] Wespestad, V., Fritz, L., Ingraham, W., Megrey, B. (2000). On relationships between cannibalism, climate variability, physical transport, and recruitment success of Bering Sea walleye Pollock (Theragra chalcogramma). *ICES Journal of Marine Science*, 57, 268–274.

[8] Cabral, H. N., Murta, A. G. (2002). The diet of blue whiting, hake, horse mackerel and mackerel off Portugal. *Journal of Applied Ichthyology*, 18(1), 14-23.

[9] Guichet, R. (1995). The diet of European hake (Merluccius merluccius) in the northern part of the Bay of Biscay. *ICES Journal of Marine Science*, 52, 21-31.

[10] May, R. M. (1984). *Exploitation of Marine Communities*. Report of the Dahlem Workshop on Exploitation of Marine Communities. Dahlem Konferenzen. Berlin: Springer-Verlag.

[11] Seijo, J. C., Defeo, O., Salas, S. (1997). *Bioeconomía pesquera. Teoría, modelación y manejo*. FAO Documento Técnico de Pesca. No. 368. Roma: FAO. [*Fisheries bio-economy. Theory, modeling, and management*. FAO Fishing Technical Issue. No. 368. Roma: FAO].

[12] FAO. (1978). *Fishery commodity situation and outlook*. Food and Agriculture Organization (FAO): Committee on Fisheries, 12th Session, COFI/78/inf. 5, 1–15.

[13] Huppert, D. (1979). Implications of multipurpose fleets and mixed stocks for control policies. *Journal of the Fisheries Research Board of Canada*, 36 (7), 845–854.

[14] Clark, C. W. (1985). *Bioeconomic Modelling of Fisheries Management*. New York: J. Wiley and Sons.

[15] Fletcher, J. J., Howitt, R. E., Johnston, W. E. (1988). Management of multipurpose heterogeneous fishing fleets under uncertainty. *Marine Resource Economics*, 4, 249–270.

[16] Anderson, L. G. (1975) Analysis of open-access commercial exploitation and maximum economic yield in biological and

technologically interdependent fisheries. *Journal of the Fisheries Research Board of Canada*, 32, 1825–1842.

[17] Mitchell, C. L. (1982). Bioeconomics of multispecies exploitation in fisheries: management implications. *Can. Spec. Publ. Fish. Aquat. Sci.*, 59, 157–162.

[18] Keddy, P. A. (1989). *Competition*. London: Chapman and Hall.

[19] Giller, P. S. (1984). *Community Structure and the Niche*. London: Chapman and Hall.

[20] Begon, M., Harper, J. L., Townsend, C. R. (1990). *Ecology. Individuals. Populations and Communities*. Boston: Blackwell Scientific Publications.

[21] Larkin, P. A. (1982). Aquaculture in North America: an assessment of future prospects. *Canadian Journal of Fisheries and Aquatic Sciences*, 39, 151–154.

[22] Majkowski, J. (1982). Usefullness and applicability of sensitivity analysis in a multispecies approach to fisheries management. In: Pauly, D. and Murphy, G. I. (ed.): Theory and Management of Tropical Fisheries. ICLARM Conference Proceedings, 9, 149–165.

[23] Sugihara, G., García, S., Gulland, J. A., Lawton, J. H., Maske, H., Paine, R. T., Platt, T., Rachor, E., Rothschild, B. J., Ursin, E. A., Zeitzschel, B. F. K. Ecosystem dynamics. In: May, R. M., [ed.], Exploitation of marine communities, report of the Dahlem Workshop on Exploitation of Marine Communities; 1984.

[24] Link, J. S. (2009). *Ecosystem-Based Fisheries Management: Conforting Tradeoffs*. New York: Cambridge University Press.

[25] Ströbele, W. J. and Wacker, H. (1991). The concept of sustainable yield in multi-species fisheries. *Ecological Modelling*, 53, 61–74.

[26] Flaaten, O. (1991). Bioeconomics of sustainable harvest of competing species. *Journal of Environmental Economics and Management*, 20, 163–180.

[27] ICES. Annual Report of the International Council of Exploration of the Sea (ICES) Advisory Committee. ICES WGHMM. www.ices.dk; 2015.

[28] ICES. Annual Report of the International Council of Exploration of the Sea (ICES) Advisory Committee. ICES WGWIDE. www.ices.dk; 2015.

[29] Sánchez, F. (1993). *Las comunidades de peces de la plataforma del Cantábrico*. Madrid: Publicación Especial del Instituto Español de Oceanografía, 13, 1993. [*The fish communities on the Cantabrian shelf.* Madrid: Special Publication of the Spanish Oceanography Institute, 13, 1993].

[30] Velasco, F. and Olaso, I. (19998). European Hake Merluccius merluccius (L., 1758) feeding in the Cantabrian Sea: seasonal, bathymetric and length variations. *Fisheries Reasearch*, 38, 33-44.
[31] Velasco F., Olaso, I., Sánchez, F. (2003). Annual variations in the prey of demersal fish in the Cantabrian Sea and their implications for food web dynamics. *ICES Marine Science Symposia*, 219, 408–410.
[32] Gonzalez, R., Olaso, I., Pereda, P. (1985). Contribución al conocimiento de la alimentación de la merluza (Merluccius merluccius L.) en la plataforma continental de Galicia y del Cantábrico. *Boletín del Instituto Español de Oceanografía*, 2, 49–60. [Contribution to the knowledge of the food of hake (Merluccius merluccius L.) in the Galician and Cantabrian continental shelf. *Bulletin of the Spanish Institute of Oceanography*, 2, 49-60].
[33] MAGRAMA. (2015). *Anuario de Estadística*. Ed. Secretaría General Técnica. Subdirección General de Estadística. Servicio de Estadísticas de la Pesca. Madrid: Ministerio de Agricultura, Alimentación y Medio Ambiente. [*Yearbook of fishing statistics*. Ed. Technical General Secretariat. General Sub-direction for Statistics. Fisheries Statistics. Madrid: Ministry of Agriculture, Food and Environment].
[34] Brown, G., Berger, B., Ikiara, M. (2005). A Predator-Prey Model with an Application to Lake Victoria Fisheries. *Marine Resource Economics*, 20, 221–247.
[35] Clark, C. W. (1976). *Mathematical Bioeconomics - the Optimal Managemet of Renewable Resources*. Sussex: J. Wiley and Sons.
[36] Clark, C. W. and Munro, G. R. (1975). The economics of fishing and modern capital theory:a simplified approach. *Journal of Environmental Economics and Management*, 5(2), 96–106.
[37] Kamien, M. and Schwartz, N. (1991). *Dynamic Optimization. The Calculus of Variations and Optimal Control in Economics and Management*. New York: North-Holland Ed.
[38] ARVI. (2014). *El sector pesquero en Vigo. Evaluación de su impacto económico*. Vigo: Cooperativa de armadores del Puerto de Vigo y Autoridad Portuaria de Vigo. [*The fishing sector in Vigo. Assessment of their economic impact*. Vigo: Ownership Association of the port of Vigo and Vigo Port Authority].
[39] MAGRAMA. (2015). *Encuesta Económica de Pesca Marítima 2014. Principales Resultados*. Ed. Secretaría General Técnica. Subdirección General de Estadística. Servicio de Estadísticas de la Pesca. Madrid: Ministerio de Agricultura, Alimentación y Medio Ambiente. [*Economic*

survey of marine fishing 2014. Main results. Ed. Technical General Secretariat. General Sub-direction for Statistics. Fisheries Statistics. Madrid: Ministry of Agriculture, Food and Environment].

BIOGRAPHICAL SKETCHES

Mª Dolores Garza-Gil received her BA degree from University of Santiago de Compostela (Spain) in 1990. She is PhD in Economics since 1995 and Professor in Department of Applied Economics in University of Vigo (Spain) since 2009. She is an expert in fisheries economics focusing on bio-economics modelling, economic evaluation and fisheries management and regulation. She has participated in several international projects and her research has been published in different journals: among others, Marine Policy, Ecological Economics, Environmental and Resources Economics, Disasters, American Journal of Agricultural Economics, Ocean and Coastal Management and Fisheries Research. Some publications in the last three years are:

- Garza; Varela (2015): "The preferences of the Spanish fishermen and their contribution on reform of the European Common Fisheries Policy," Ocean and Coastal Management, 116, 291-299.

- Garza; Amigo; Surís; Varela (2015): "Perceptions on incentives for compliance with regulation: The case of Spanish fishermen in the Atlantic," Fisheries Research, 170, 30-38.

- Garza-Gil, M.D.; Varela-Lafuente, M.; Caballero-Miguez, G.; Torralba-Cano, J. (2015): "A study on economic impact on the European Sardine Fishery due to continued global warming," Bharat R. Singh, Global Warming. Causes, Impacts and Remedies, 115-136. Intech.

- Caballero; Varela; Garza (2015): "From effort-control to an ITQ system in the Grand Sole Fishing Grounds: Evolution and new evidence from a case-study," FAO Technical Paper, Chapter 8, en SQUIRES, D., M. Maunder, N. Vestergaard, V. Restrepo, R. Metzner, S.Herrick, R. Hannesson, I. del Valle, and P. Anderson (editors): Effort Rights in Fisheries Management: General Principles and Case Studies from Around the World. 2015. FAO Fisheries and Aquaculture Proceedings P34. Rome: Food and Agriculture Organization of the United Nations. Forthcoming.

- Surís; Garza; Varela (2014): "Socio-economic quantification of fishing in a European urban area: The case of Vigo," Marine Policy, 43, 347-358

- Caballero; Varela; Garza (2014): "Institutional change, fishing rights and governance mechanisms: The dynamics of the Spanish 300 fleet on the Grand Sole fishing grounds," Marine Policy, 44, 465-472.

Marcos I. Pérez-Pérez is Ph.D in Economics since 2016 and researcher in Department of Applied Economics in University of Vigo, Spain, since 2009. He is an expert in fisheries economics. He has participated in several international projects and his research has been published in different national books and journals. His last publication in an international journal is "Pulse Fishing and stock uncertainty," Environmental Resource Economics, 59(2).

Manuel Varela-Lafuente is Ph.D in Economics since 1982 and Professor in Department of Applied Economics in University of Vigo (Spain), since 1992. He was director of the Technological Center of the Sea (CETMAR) in Vigo between 2005 and 2007. He is an expert in natural resource economics, especially in fisheries economics focusing on economic evaluation and fisheries management and regulation. He has participated in several international projects and his research has been published in different journals: among others, Marine Policy, Fisheries Research, The Journal of Disasters Studies, Policy and Management, American Journal of Agricultural Economics, Ocean and Coastal Management. Some publications in the last three years are:

- Garza; Varela (2015): "The preferences of the Spanish fishermen and their contribution on reform of the European Common Fisheries Policy," Ocean and Coastal Management, 116, 291-299.
- Garza; Amigo; Surís; Varela (2015): "Perceptions on incentives for compliance with regulation: The case of Spanish fishermen in the Atlantic," Fisheries Research, 170, 30-38.
- Garza-Gil, M.D.; Varela-Lafuente, M.; Caballero-Miguez, G.; Torralba-Cano, J. (2015): "A study on economic impact on the European Sardine Fishery due to continued global warming," Bharat R. Singh, Global Warming. Causes, Impacts and Remedies, 115-136. Intech.
- Caballero; Varela; Garza (2015): "From effort-control to an ITQ system in the Grand Sole Fishing Grounds: Evolution and new evidence from a case-study," FAO Technical Paper, Chapter 8, en SQUIRES, D., M. Maunder, N. Vestergaard, V. Restrepo, R. Metzner, S.Herrick, R. Hannesson, I. del Valle, and P. Anderson (editors): Effort Rights in Fisheries Management: General Principles and Case Studies from Around the World. 2015. FAO Fisheries and

Aquaculture Proceedings P34. Rome: Food and Agriculture Organization of the United Nations. Forthcoming.

- Surís; Garza; Varela (2014): "Socio-economic quantification of fishing in a European urban area: The case of Vigo," Marine Policy, 43, 347-358
- Caballero; Varela; Garza (2014): "Institutional change, fishing rights and governance mechanisms: The dynamics of the Spanish 300 fleet on the Grand Sole fishing grounds," Marine Policy, 44, 465-472.

Juan C. Surís-Regueiro is Professor in Department of Applied Economics in University of Vigo (Spain) since 2002. He is an expert in fisheries economics focusing on economic evaluation and fisheries management and regulation. He was President of the port of Marín (in Galicia, Spain) between 2005 and 2009. He has participated in several international projects and his research has been published in different journals: among others, Marine Policy, Disasters, Fisheries Research, and Ocean and Coastal Management. Some publications in the last three years are:

- Garza; Amigo; Surís; Varela (2015): "Perceptions on incentives for compliance with regulation: The case of Spanish fishermen in the Atlantic," Fisheries Research, 170, 30-38.
- Surís; Garza; Varela (2014): "Socio-economic quantification of fishing in a European urban area: The case of Vigo," Marine Policy, 43, 347-358
- Caballero; Varela; Garza (2014): "Institutional change, fishing rights and governance mechanisms: The dynamics of the Spanish 300 fleet on the Grand Sole fishing grounds," Marine Policy, 44, 465-472.
- Surís; Santiago (2014): "Characterization of fisheries dependence in Galicia (Spain)," Marine Policy, 47, 99-109.

INDEX

D

E

F

G

H

I

J

K

L

M